MATERIALS SCIENCE AND TECHNOLOGIES

SEMICONDUCTOR MATERIALS FOR GAS SENSORS

MATERIALS SCIENCE AND TECHNOLOGIES

Additional books in this series can be found on Nova's website under the Series tab.

Additional e-books in this series can be found on Nova's website under the e-book tab.

MATERIALS SCIENCE AND TECHNOLOGIES

SEMICONDUCTOR MATERIALS FOR GAS SENSORS

VALENTYN SMYNTYNA

nova publishers
New York

For permission to use material from this book please contact us:
Telephone 631-231-7269; Fax 631-231-8175
Web Site: http://www.novapublishers.com

NOTICE TO THE READER

Additional color graphics may be available in the e-book version of this book.

Library of Congress Cataloging-in-Publication Data

Smyntyna, V. A. (Valentyn Andriiovych), 1948-
Semiconductor materials for gas sensors / Valentyn Smyntyna (Faculty of Physics, Odessa National I.I. Mechnikov University, Odessa, Ukraine).
pages cm
Includes bibliographical references and index.
ISBN: 978-1-62808-458-0 (hardcover)
1. Gas detectors--Materials. 2. Semiconductors. I. Title.
TP754.S59 2013
621.3815'2--dc23

2013025323

Published by Nova Science Publishers, Inc. † New York

CONTENTS

INTRODUCTION

Research in the field of sensor physics and technology is one of the priorities at the research centers around the world. This is being demonstrated by regularly scheduled international congresses, conferences and exhibitions, such as TRANSDUSERS, SENSORS, EURUSENSORS, BIOSENSORS, etc. Five international conferences of the series "Sensors Electronics and Microsystems Technology" are already held, where scientists from more than 40 countries participated. Interest in this area is due, in particular, to the need of monitoring the gas composition of the atmosphere in general, toxic air emissions from various industrial processes and to control of the composition of air in the habitat and human activity.

Widely used in gas analysis, the semiconductor sensors have a number of advantages: they are resistant to aggressive media, temperature, have fast response, simple design and are generally cheaper as to the manufacturing technology.

Sensor properties depend on the nature of semiconductor material, its alloying composition, dispersion source material and on the technological way of production. The above mentioned parameters and details determine the structure and conditions of the surface of a semiconductor, which, in their turn, influence adsorption properties of the surface.

Basic information parameters of semiconductor gas sensors are the characteristics that vary considerably in the course of adsorption processes (conductivity, surface potential etc.). The most significant changes of recorded parameters could be easily observed in metal oxide semiconductors, as well as in semiconductors of A_2B_6 group and other semiconductors. Sufficiently serious constraint to the use of these materials as gas sensors is their low selectivity.

Band model, in the framework of which the mechanisms are interpreted through changing the conductivity of semiconductors during their interaction with different gases is not sufficient for a complete understanding of the mechanism of surface electron-molecular processes. Significant role in these processes is played by the chemical composition, surface structure of the semiconductor and the local interaction between the atom and the surface of the semiconductor alien gas particle. Therefore, it is important to create a quantum-mechanical model of physical and chemical processes on the surfaces of semiconductors as a theoretical basis for the development of gas sensors. To achieve this goal, one should carry out the simulation of electron-molecular phenomena on the surface of semiconductor films, to investigate the mechanism of adsorption sensitivity of semiconductor films as well as the adsorption kinetic characteristics and features of the sensitivity degradation of the adsorption of the studied sensors.

Alongside with metal oxide semiconductor films and layers of porous silicon, films of A_2B_6 group are considered as the model compounds, because the materials in this group have high gas sensitivity due to the chemical composition of the surface and the possibility of varying the position of the Fermi level in a wide range. These characteristics make the contact A_2B_6 irreplaceable model systems in the study of adsorption and catalytic processes on semiconductor surfaces.

Tasking into account the long-term operation of the sensors, there are irreversible changes in the stoichiometric composition of the surface, which affects the electrical and adsorption-sensitive parameters of the sample, such as resistance and sensitivity of adsorption, respectively.

In this monograph, the readers could find some basic concepts and patterns of electron-molecular interactions on the surface of the semiconductor, analyzes the processes of formation of the adsorption properties of the surface during production and further processing of semiconductor materials. Experimental results of a number of semiconductor gas sensors to monitor changes in the concentrations of certain gases in the atmosphere are also are considered in this book.

This book is logical continuation of issues discussed earlier in my book published in 2013 by NOVA Press (Valentyn Smyntyna. *Electron and Molecular Phenomena on the Surface of Semiconductors),* which systematically examines results of investigation of electronic and molecular processes on the surfaces of semiconductors, taking place at their interaction with particles of gas environment or in the course of superficial alloying by atoms of metals. The main subject of the book is analysis of interaction of semiconductors with foreign atoms and molecules from gas environment and from beams of elements, bombarding a surface.

Current monograph can be useful for researchers working both theory and experimental study of the problem of electron-molecular phenomena on the surface of semiconductor gas sensors, as well as in the field of their practical application. Separate sections and paragraphs are applicable in teaching in universities and working with graduate and post-graduate students and masters. It is assumed that the reader is sufficiently informed about the issues and challenges in this field of research.

Valentyn Smyntyna

ACKNOWLEDGMENTS

In 1970 for the first time I was surprised to find a dramatic degradation of the film resistance of cadmium selenide film at the first contact with the air immediately after it's been prepared in a vacuum. This happy occasion has become for me the starting point of interesting work, which continues today. Over the years more and more researchers studying the interaction of gas atmosphere and biological objects (our recent subject of studies) with the surface of semiconductors were concentrating around of me on the base of my chair of experimental physics, the laboratory of molecular and electronic processes on semiconductor surfaces and physical and technical center. It is a pleasure to thank them all for their enthusiasm and perseverance in work, as well as for their ingenuity in experiments and in the interpretation of their results. Together we have prepared dozens of their PhD and Doctor of Sciences (Habilitated) theses, which they successfully defended, hundreds of our joint presentations at international forums and publications in the international peer-reviewed journals representing results of our joint experiments. I am deeply grateful to my colleagues who continue working with me as well as for those who are now continuing their careers in laboratories around the world.

Preparation of this monograph for publication is completed largely due to the active and useful help of professors O. Smyntyna, O. Fedchuk and Yu. Nitsuk, employee Yu. Novosolova and graduate student E. Britavsky to whom I am sincerely grateful for their tremendous work.

I highly appreciate results of our scientific cooperation with colleagues from universities and research institutes in Italy, Finland, Germany, Latvia, Poland, Slovakia, Sweden, Russia and Ukraine in frames of series of European projects as well as on the base of bilateral agreements.

I address my deep gratitude to Current and Honorary member of the EUROSENSORS International Steering Committee for their collegial stimulation of research in the field of sensors.

Dynamic equilibrium with the environment is impossible to achieve without stability in personal life. I heartily thank my wife Gianna and daughter Olena for my excellent ground state and appreciate their understanding, patience and selfless help.

Valentyn Smyntyna

Chapter 1

FEATURES OF THE ELECTRON-MOLECULAR INTERACTIONS ON THE SURFACE OF THE SEMICONDUCTOR IN THE CLASSICAL AND ELECTRONIC THEORY OF ADSORPTION-DESORPTION PROCESSES

ABSTRACT

At the beginning of the first chapter briefly describes the main provisions of the classical theory of adsorption on the basis of the Langmuir equation and assumptions. Some important features of the Langmuir adsorption kinetics and limits of its applications due to the provisions of the theory are examined.

The main consequences of the electron theory of chemisorption are discussed. The nature of the physical and chemical adsorption is taken into account. The possibilities of activated and non-activated adsorption are examined.

Electronic transitions in chemisorption on semiconductor surfaces are described by the Fermi level. The Fermi level is used as the primary regulator of the process parameters.

On the basis of the electron theory of chemisorption the impact on the chemisorption processes of crystal size, the change in the surface charge, surface potential, work function, and the electrical conductivity of the semiconductor are examined.

In the final section of this some of the key features of chemisorption are analyzed and compared with the local and collective interactions.

1.1. Summary of the Theory of Adsorption Basics

The process of adsorption, the contact of the gas environment with the solid surface, filled with gas molecules and atoms. This process occurs at a constant temperature for as long as between the surface of the solid and gaseous media, the adsorption equilibrium is established. [1] The kinetics of the adsorption process can be described by:

$$N=N(t) \tag{1.1}$$

where N - is the concentration of adsorbed particles on the surface, t - time.

If the adsorption equilibrium number of gas molecules N is steady, depending on the pressure P and temperature T, the dependence is presented by:

$$N=N(P, T). \tag{1.2}$$

this equation (1.2) is called the equation of state. If in the equation of state T = *const*, then we get the equation of the isotherm

$$N=N(P)_{T=const}.$$

At P = *const* (1.2) becomes the equation of the isobar

$$N=N(T)_{P=const}.$$

Classical theory of adsorption, proposed by Langmuir, is based on several key assumptions [2].

1. Adsorption occurs on the individual adsorption centers. The nature of these centers may be unknown. Each adsorption center can hold only one gas particle. On the surface adsorption sites could be only of one type. These centers have the same energy of the gas particles interaction with the centers. Surface having these properties is called a uniform energy surface.
2. The bond strength of the molecule with the adsorption center is determined only by the nature of the molecule and the nature of the adsorption sites and does not depend on the presence or absence of

molecules in adjacent adsorption sites. In other words, the adsorbed molecules do not interact with each other.

3. The number of adsorption centers on the surface is constant. Its value characterizes this area and is caused by surface features biographical. The total value of the number of adsorption centers on the surface does not depend on temperature and surface coverage by adsorbed molecules.
4. The binding energy between the center of the adsorption and the adsorbed particle is unique characteristic center of the adsorption center and the adsorbed particles as well. This means that the separation of the center of adsorption and adsorbed particle can bind only in one specific way. Over the lifetime of the adsorbed molecules on a dedicated adsorption state the nature and strength of the bonding it to the center of adsorption does not change.

1.2. Langmuir Adsorption Kinetics

Let N^* - be the concentration of adsorption centers, and N - concentration of adsorbed particles and P - pressure. Then, according to Langmuir, one could write the equation of the adsorption kinetics of the particles of the class:

$$\frac{dN}{dt} = \alpha P(N^* - N) - \beta N. \tag{1.3}$$

In equation (1.3), being called the Langmuir equation, we have:

$$\alpha = \frac{vs}{\sqrt{2\pi MT}} \tag{1.4}$$

$$\beta = \theta \exp\left(-\frac{E_\delta}{kT}\right) \tag{1.5}$$

The first term on the right side of (1.3) , is the number of molecules approaching per 1 sec per $1m^2$ surface, and the second term describes the number of molecules leaving the surface of $1m^2$ per 1 sec. M - mass of the adsorbed molecules, s - its effective area, ν - the probability that a gas molecule that hit the center of the adsorption, which assumed to be fixed, β - probability of desorption of adsorbed molecules per unit time, $\tau = 1/\beta$ - average lifetime of the molecules in the adsorbed state, E_δ - binding energy of the adsorbed particles with an adsorption center.

In some cases, the kinetics of adsorption is described by Roginsky - Zeldovich equation [3]:

$$\frac{dN}{dt} = C\exp(-\gamma N) \tag{1.6}$$

where $\gamma > 0$. If $\gamma C = 1/t_0$, one could get:

$$N(t) = \frac{1}{\gamma}\ln\left(1 + \frac{t}{t_0}\right) \tag{1.7}$$

Kinetics also may be subject to the following equation:

$$\frac{dN}{dt} = C\exp(-\gamma N^2) \tag{1.8}$$

In some cases, the kinetics is described by the empirical Benhem`s equation:

$$N(t) = Ct^n \tag{1.9}$$

In equation (1.9) $n < 1$, $C = const$.

Let us consider some particular solutions of Langmuir (1.3) and the general solution of this equation. In the initial moments of interaction between adsorbed particles with a solid surface, N is sufficiently small. In this case we can ignore the desorption term in (1.3), and the value of N as compared to N^*. Then (1.3) takes the following form:

$$\frac{dN}{dt} = \alpha P N^* \tag{1.10}$$

Its solution is presented by the expression:

$$N = N^{*}\frac{t}{t_0}, \ (t_0 = 1/\alpha P). \tag{1.11}$$

Equation (1.11) has a physical sense, only before $N = N^*$, that is, before $t = t_0$. Hence, (1.11) can be considered only in the interval $0 < t < t_0$ (Figure 1.1).

Let the concentration of N be such, that the desorption can still be neglected. However, changes of N in the term $(N^* - N)$ should be considered. Then from (1.3):

$$\frac{dN}{dt} = \alpha P(N^* - N) \tag{1.12}$$

one could get

$$N(t) = N^*\left[1 - \exp(-\alpha P t)\right] \tag{1.13}$$

If we denote $\alpha P = 1/t_{01}$, then

$$N(t) = N^*\left[1 - \exp\left(-\frac{t}{t_{01}}\right)\right] \tag{1.14}$$

Equation (1.14) has a physical meaning as long as $N < N^*$, that is, in the time interval $0 < t < \infty$ (Figure 1.1).

If adsorption is being advanced, when the value of N is relatively large and desorption cannot be neglected, it is necessary to solve the complete Langmuir equation (1.3).

Transforming (1.3) the right part, one has:

$$\frac{dN}{dt} = \alpha P N^* - N(\alpha P + \beta) \tag{1.15}$$

Dividing the left and right by the part of $(\alpha P + \beta)$:

$$\frac{1}{\alpha P+\beta}\frac{dN}{dt}=\frac{\alpha P}{\alpha P+\beta}N^{*}-N \tag{1.16}$$

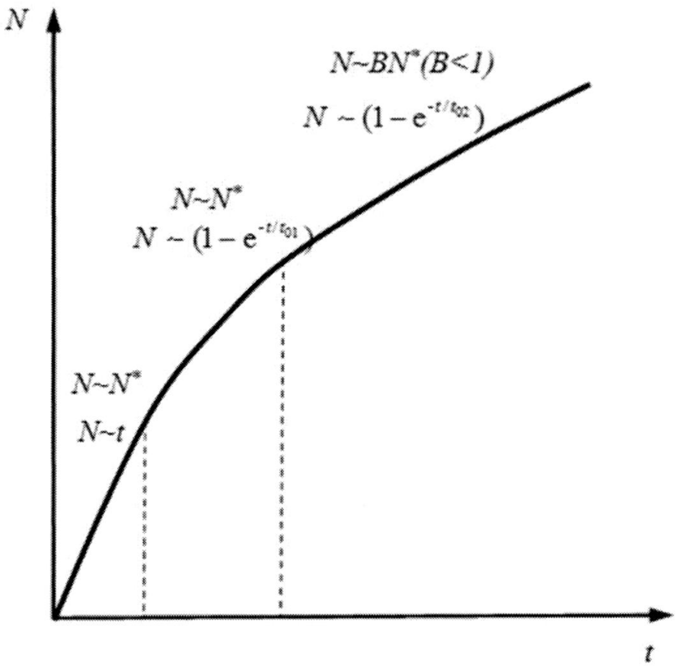

Figure 1.1. Langmuir adsorption kinetics. (see, please, text for details).

Transformed coefficients of the first term on the right side are:

$$\frac{\alpha P+\beta}{\alpha P}=1+\frac{\beta}{\alpha P}$$

and denoting

$$\frac{\beta}{\alpha}=b \tag{1.17}$$

and after integration with the initial conditions $N = 0$, $t = 0$, concerning the notation:

$$a = \frac{1}{t_{02}}; \frac{1}{1+\frac{b}{P}} = B \tag{1.18}$$

one obtains the solution of (1.3) in general form (see Figure 1.1)

$$N(t) = BN^{*}\left[1 - \exp(-\frac{t}{t_{02}})\right], \tag{1.19}$$

$$\left(a = \frac{1}{t_{02}} = \alpha P + \beta,\ t_{02} = \frac{1}{\alpha P + \beta}\right).$$

From these equations, one could see that important part and the kinetics of adsorption are the time constants. Let us consider their physical content.

We return to equation (1.14):

$$N(t) = N^{*}\left[1 - \exp\left(-\frac{t}{t_{01}}\right)\right]$$

At $t = t_{01}$

$$N = N^{*}\left(1 - \frac{1}{2,7}\right) = 0,63N^{*} \tag{1.20}$$

Consequently, t_{01} is the time at which the concentration of chemisorbed particles becomes equal to 0.63 N *.

According to the Weitz limit:

$$N_{max} = (10^{-2}–10^{-3})N^{*}. \tag{1.21}$$

Comparing (1.20) and (1.21), we can see that together with the underlying equation, the (1.12) is a rather crude approximation. This follows from the fact, that even at $t = t_{01}$ obtained with the help of the values of N in (10 - 100) times higher than the maximum allowable.

However, equation (1.12) is more correct than (1.11), since the solution of the latter leads to physical nonsense, which consists in the fact that at $t \to \infty$ one obtains $N \to \infty$. It follows that the kinetics (1.11) is applicable only in the very early stages of adsorption. The kinetics of adsorption (1.14), can be used in the interval $0 < t < (10^{-2} - 10^{-3})\, t_{01}$.

We now turn to t_{02}. From the complete solutions of Langmuir equation (1.19), one could see that at $t = t_{02}$:

$$N = 0.63\, B\, N^*. \tag{1.22}$$

while at:

$$t \to \infty,\ N \to B\, N^*. \tag{1.23}$$

From (1.23) it follows that, according to the Weitz limit, $B \approx 10^3$. Thus, the solution of the complete Langmuir equation most closely matches the real processes on solid surfaces. This decision may be true in almost any experimentally observed time interval, if the parameter $B < 10^{-3}$. When $t \to \infty$, adsorption rushes to equilibrium desorption process, i.e., to the thermodynamic equilibrium. From the condition of thermodynamic equilibrium, according to (1.22), it follows that

$$N = B\, N^*$$

and

$$N = \frac{N^*}{1 + \frac{b}{P}} \tag{1.24}$$

or

$$N = N^* \frac{b}{\frac{P}{b} + 1} \tag{1.25}$$

In (1.25) $b = \beta/\alpha$. Equation (1.25) is the Langmuir isotherm.

At low pressures, $P << b$ and (1.25) is obtained in the form:

$$N = N^* \frac{P}{b} \tag{1.26}$$

Expression (1.26) represents the Henry isotherm. The area in which the linear dependence of $N \sim P$, called the domain of Henry [4].

The above dependencies are not always observed experimentally. In some cases, it may be detected Freidlich isotherm [4]

$$N = CP^{n}, \ (C, n \text{ — } const). \tag{1.27}$$

Sometimes there is a logarithmic isotherm

$$N = C \ln (P/P_0), \ (C, P_0 \text{ — } const) \tag{1.28}$$

Suppose that after the adsorption equilibrium between the surface and the gas phase pressure of the gas atmosphere is significantly reduced. In this case, the previously adsorbed gas particles begin to leave the surface of the solid, that is, developing the desorption of gas molecules and atoms. As the part of the Langmuir theory, this process is described by (1.3), in which, however, we must now use $P = 0$.

Integrating (1.3) with the assumptions made, and using the initial conditions of $N = N_0$ at $t = 0$ (N_0 - the equilibrium value of the concentration of adsorbed particles), we obtain the concentration of the desorbed particles N_{des}:

$$N_{des.} = N_0 \exp(-\beta t). \tag{1.29}$$

According to (1.29) and (1.5), the desorption occurs more rapidly at higher temperatures. However, at high temperatures, the adsorbed particles could not always be completely removed from the surface. Instead, the molecules and atoms are adsorbed irreversibly or diffuse into the bulk of the solid.

1.3. Specificity of Physical and Chemical Adsorption of Gas Particles on Semiconductor Surfaces

In the framework of the electron theory of adsorption, there are two main forms of adsorption of particles on a solid surface: physical adsorption and chemical adsorption (or chemisorption) [4].

The physical and chemical adsorption types differ due to the nature of the forces responsible for the process and ensure the retention of the adsorbed particles on the surface of a solid. In this case, the interaction forces between the adsorbed particle and a solid surface may be of different origin and nature.

If the interaction between the adsorbed particle and solid surface is provided by the Van-der-Waals, electrostatic polarization forces or the forces of electrostatic image, there is a physical adsorption.

The forces of interaction can have a chemical nature, i.e., to be the exchange type. In this case, we have the chemical adsorption (chemisorption) and the adsorption complex adsorbs the particle - the center is a of the chemical adsorption type or quasi-chemical connection. This complex interaction between its components is provided by covalent forces. In some cases, the covalent interactions are, to some extent, supplemented by ionic interactions.

In Figure 1.2 shows the adsorption curve, which is the energy dependence of W on the distance r between the surface of the adsorbent and adsorbate (q - the energy of interaction in the adsorption complex, r_0 - the equilibrium distance between the components of the adsorption complex; adsorbent - the material on which the adsorption; adsorbate - adsorbed particles)

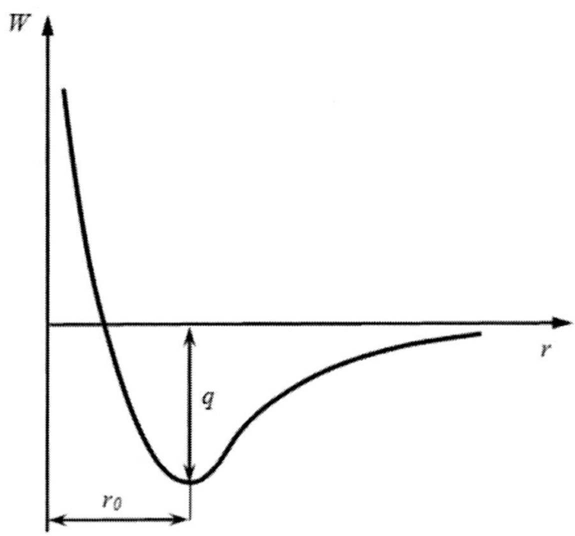

Figure 1.2. Dependence of the interaction potential energy W on the distance r between the particles. (for explanations, see, please, the text).

Chemical adsorption is characterized by a much larger binding energy q and significantly lower values of r_0, than physical adsorption. It follows that in chemisorption act, the chemisorbed particle binds to the center of the adsorption on the surface much stronger as compared to the particle in the state of physical adsorption.

In physical adsorption energy q of a particle with the center adsorption on a solid surface is approximately (0.01 - 0.1) *eV*. In the case of chemisorption, the value q is about 1 *eV*.

Physically adsorbed particle and solid can be considered as two independent systems, the interaction between them should be analyzed in the quantum theory approximation of small perturbations.

Chemisorbed particle and solid lattice compose single quantum-mechanical system, and should be considered as a whole, i.e., the adsorption complex is a chemical compound of adsorbed particles and solid.

1.4. Activated Adsorption on Semiconductor Surfaces

For implementation of the chemisorption, in many cases, it is necessary to supply some energy, in other words to activate the process. It is necessary to distinguish between the activated adsorption of chemical adsorption.

Activated adsorption is a process carried out after pre-excitation (activation) of the adsorption of the complex or the system as a whole. For the activated adsorption requires up-front costs of energy - activation, which in result of the act of adsorption, returns [4].

Despite the fact that the chemical adsorption requires activation, the supply of chemisorption's activation energy is not always a necessary criterion for its implementation. If the substitution reaction always occurs as a result of energy costs:

$$AB + C \rightarrow AC + B,$$

while a coupling reaction in the form:

$$A + B \rightarrow AB$$

could occur without activation, and, on the contrary, for its implementation, could release energy.

Thus, the presence or absence of the activation energy of adsorption can not clearly indicate the nature of the adsorption forces.

Activated adsorption is fundamentally different from other types of adsorption kinetics. In the absence of activation energy (non-activated adsorption), the adsorption is very rapid, so that equilibrium is established in the system almost instantly. In the case of activated adsorption process could be quite long and runs faster with the temperature increase.

At the initial stages of adsorption, there is little coverage of the surface by adsorbed particles, according to (1.3), the adsorption rate can be expressed as

$$\frac{dN}{dt} = v\frac{sN^*}{\sqrt{2\pi MkT}} \tag{1.30}$$

The difference between the activated and non-activated adsorption is reduced to changes, contained in the frequency factor v. For non-activated adsorption, it is assumed that $v = 1$. In this case, from (1.30) with increasing temperature, the rate of adsorption is inversely proportional to the square root of the temperature. If there is activated adsorption, the

$$v \sim exp\ (-E/kT), \tag{1.31}$$

(E - activation energy). Thanks to (1.31), the adsorption rate increases exponentially with increasing temperature. The presence of such an exponential factor is a characteristic feature of activated adsorption. In the theory of activated adsorption it is considered the influence of activation barrier height on the flow of adsorption as a physical cause, which requires certain energy for the occurrence of adsorption. The height of the energy barrier can be different due to various reasons.

One of the mechanisms of activation barrier was proposed by Lennard-Jones [5]. It was showed that the activation barrier can occur if the adsorption of the molecule requires its dissociation into atoms (Figure 1.3). Dissociation into atoms is shown in Figure 1.3 as the transition from the position of the atoms x_1 in the position x_2. On the power W dependence on the distance x, of the hydrogen molecules H_2 from a solid surface (Figure 1.3, b), it is observed a shallow minimum on the curve 1 (at $x = x_1$), which corresponds to the chemisorption of hydrogen atoms, which indicates the implementation of the dissociation of the molecule H_2. Therefore, the transition from curve 1 to curve 2 is accompanied by dissociation of the hydrogen molecule. For the transition from a state described by curve 1, to the state corresponding to curve 2, in order to overcome the energy barrier, which height corresponds to the intersection of curves 1 and 2 (x_0). The height of the barrier E is the activation energy of the considered process.

The height of the activation barrier E in Figure 1.3, b can differ from the true values of the activation energy of adsorption. These differences depend on several factors. The state of the system, described by curves 1 and 2 in Figure 1.3, correspond to different values of y, and, therefore, the value of E is not quite correct unambiguously interpreted as the height of the activation barrier.

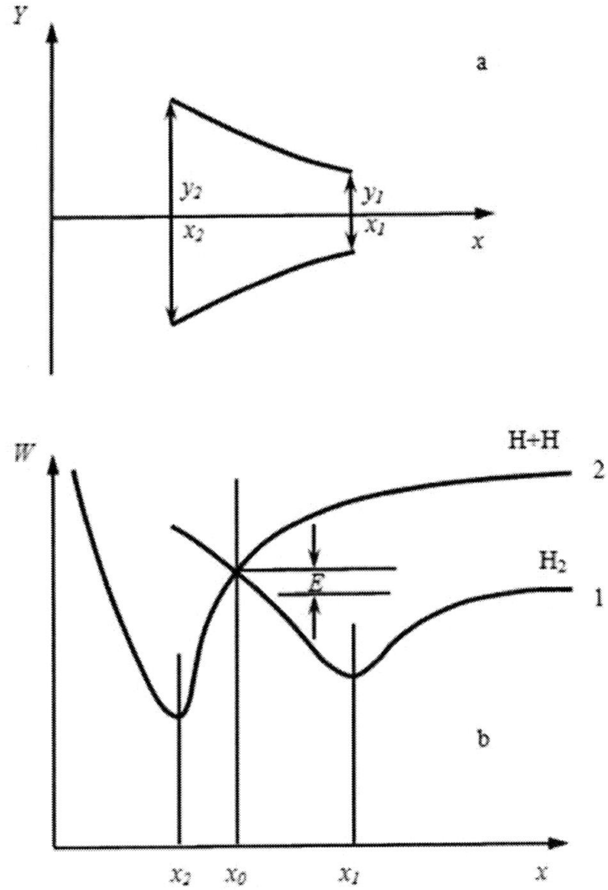

Figure 1.3. Chemisorption potential curves for the case of di-atomic molecule. (see, please, explanations in the text).

According to (1.30) $N^* = const,\ v \sim exp\ (-E/kT)$. (1.32)

The number of adsorption sites is constant and independent of temperature and determined only by the nature of the solid. It is not always confirmed experimentally, and there are cases, where one has:

$$N^* \sim exp\ (-E/kT),\ v = const. \quad (1.33)$$

If the conditions (1.33) are fulfilled, one could also see all of the features of activated adsorption, but they will be determined by the nature of adsorption sites. The activation energy, in this case, characterizes the energy necessary for the formation of adsorption centers.

On solid surfaces could be two types of adsorbed particles, simultaneously, which, according to (1.33), are characterized by two different energies E_1 and E_2. In this case, the activation energy depends not only on the nature of adsorption centers, but also the nature of the adsorbed particles.

In general, for the same type of adsorbed particles could be written as

$$N^* \sim exp\ (-E_1/kT),\ v \sim exp\ (-E_2/kT) \quad (1.34)$$

(E_1 - the activation energy of adsorption centers' formation, E_2 – the activation energy of chemisorption of the particles). Full activation energy of the adsorption process in this case is the sum of

$$E = E_1 + E_2. \quad (1.35)$$

The presence of an activation barrier for adsorption curve is not always a clear indication of the chemical nature of the adsorption forces. Such a barrier can occur during physical adsorption in the case that the forces acting between the adsorbed particles are repulsive. Indeed, in the presence of repulsive forces, on each new adsorbed particle there are two opposite forces acting. One of them is due to the forces acting on the particle for adsorbed atoms on the surface of the solid, which it attracts. The repulsive type of the force is due to the action of repulsive inter-particle' force for the particles, already adsorbed on the surface. The relationship between the forces, acting on the particle approaching surface, which in some cases could lead to the formation of an energy barrier. One of cases is due to the forces acting on the particle from adsorbed atoms on the surface of the solid, which is of the attracting type. Repulsive force is due to the action of repulsive particles already adsorbed on the surface. The relationship between the forces, acting on the approaching particle to the surface, which, in some cases, can lead to the formation of an energy barrier.

Consider the surface, previously partially filled by physically adsorbed molecules. Let this new surface is approaching by the same kind of particles. In this case, the interaction energy *W(r)* between the particle and the surface is described by the curve in Figure 1.4, a [1].

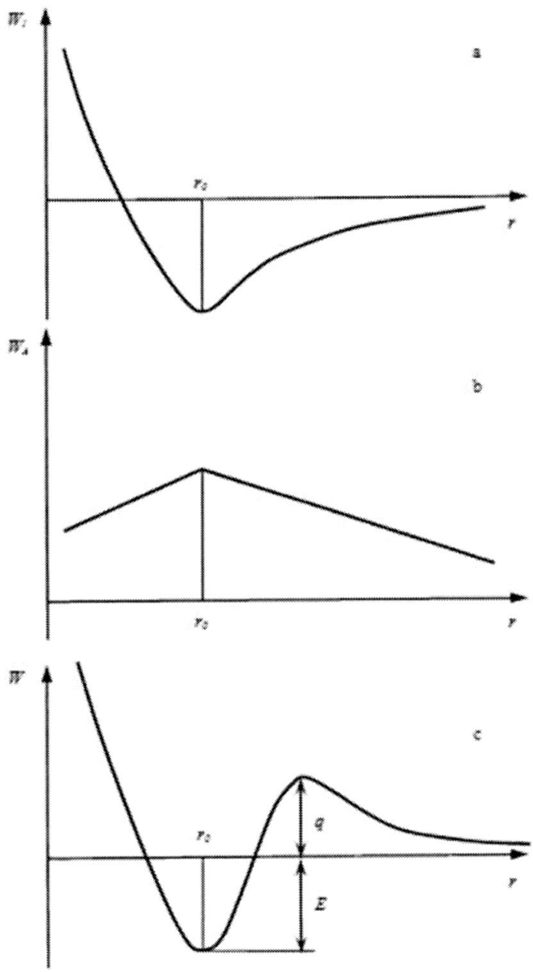

Figure 1.4. Distance dependence of the interaction energy of the adsorbed particles with a solid surface (a), with all the previously adsorbed particles (b) and the change in the total interaction energy (c).

The interaction of adsorbed particles previously adsorbed particles is shown in Figure 1.4, b. The addition of the curves in Figure 1.4, (a and b) leads to the curve shown in Figure 1.4, c. As a result of complex interaction of adsorbed particles with a solid surface and the previously adsorbed particles, it is obtained, that the higher the surface coverage by adsorbed particles, the higher is the maximum of the curve in Figure 1.4, b and the higher the barrier height E, and the smaller the depth of energy well q in Figure 1.4, c. Therefore, the activation barrier may occur in the case of physical adsorption as a result of interactions between the adsorbed particles.

1.5. Electronic Transitions in Chemisorption Process on Semiconductor Surfaces

The electronic theory of adsorption "electronic transfer" means the transfer of an electron in the crystal lattice of the solid particles chemisorbed on the surface. In this case, it is assumed that the electrons, belonging to the chemisorbed particle, during its interaction with the solid leaves the particle and is owned by the solid. The probability of finding an electron in a specified volume of a solid, is non-zero and zero outside it. [4]

There is another case in which the electrons belonging to the crystal structure of a solid, in the act of its interaction with the chemisorbed particle is localized to this bit. In the latter case, the wave function of the electron has a maximum as a function of adsorbed particles' plane and decreases with distance from the particle in the direction of the solid.

Adsorbed particle and solid can form several types of connections, which include:

a) weak form of chemisorption,
b) strong donor connection
c) strong acceptor bond.

It is assumed that the weak form of chemisorption is a "tightening" of the electron from the chemisorbed particles into the volume of the crystal lattice. Greater or lesser degree of tightening of the electron in the lattice depends on the nature of the solid and the adsorbed particles. It does not result in a complete transfer of an electron from an adsorbed particle solid. Electron continues largely "own" adsorbed particles and is localized near it [4].

In the case of a strong donor communications, the adsorbed particles act as electron donor with respect to the solid. Electron is completely absorbed by the particle crystal lattice of the solid, and the wave function of the electron is "smeared" throughout the crystal.

Formation of a strong acceptor bond is accompanied by a complete localization of the electron crystal around the adsorbed particles.

The concept of the electronic transition, therefore, is rather vague in the electronic theory of chemisorption due to the fact that it is interpreted on the basis of geometric representations of electron transport.

In the framework of quantum mechanical concepts of electronic transitions in chemisorption, are considered the processes of electron transitions from one energy level to another. These transitions change the state of the electrons in such a way, that the final states are different from the original distribution in space and energy, and this is reflected in the wave functions of the electrons. In this case, the electronic transitions are treated as transitions of electrons, for example, from the conduction band to the local energy levels created by the adsorbed particle in the gap on the surface of a solid.

1.6. The Fermi Level in Chemisorption Process

The position of the Fermi level is determined by the concentration of chemisorbed particles on a solid surface, i.e., the adsorption capacity of the surface. As far as the distance of the Fermi level from the bottom of the conduction band determines the adsorption capacity of the surface, the chemisorption of acceptor particles decreases while the adsorption capacity of donor particles increases.

Changing the position of the Fermi level in the band gap, also determine the charge state of the adsorbed particles, and hence the total amount of charge being localized on the surface of a solid. Moving down the Fermi level, from the bottom of the conduction band, one causes the release of the negative charge, accumulated on the surface as a result of adsorption, i.e. reduces the likelihood of adsorption of acceptor particles. At the same time, the share of positive charge, localized on the surface, indicates an increase of the probability of adsorption of particles of the donor type. In this case, it is clear that the position of the Fermi level in the band gap of a solid depends on the relative abundance of different forms of chemisorption. Thus, the Fermi level is not only determined by the concentration of electrons and holes on the surface of a solid, but also acts as a regulator of some of its adsorption properties [4, 6].

However, the position of the Fermi level depends on the concentration and nature of particles adsorbed on the surface of the crystal. For example, the chemisorption of acceptor particles accumulated on the surface of the negative charge and thereby cause the removal of the Fermi level of the conduction band due to the fact that in this case, a stop-band bending (up) takes place on the surface.

The observed mutual dependence between the position of the Fermi level and the degree of surface coverage by adsorbed particles creates difficulties in the analysis of the adsorption processes in the electron theory of adsorption.

Suppose that, as a result of chemisorption on the surface, changes a balance between the capture of electrons from the solid to the surface levels by the adsorbed particles of one sort and reverse transitions of electrons in the conduction band on these levels. Consider a solid body with *n*-type conductivity, on the surface of which are adsorbed the particles of acceptor type. Let us denote the total concentration of adsorbed particles by *N*, and their concentration in the neutral (physical adsorption) and negatively charged (chemisorption) states - respectively, N^0 and N^-. Then, the relative proportion of particles in a state of physical A^0 and chemical A^-adsorption can be represented as follows [4, 6]

$$A^0 = \frac{N^0}{N}, A^- = \frac{N^-}{N} \quad (1.36)$$

On the band diagram (Figure 1.5), the particles in number N^0 meet unfilled electron E_1 levels above the Fermi level, and the corresponding N^-particles - E_2 levels are below the Fermi level.

From (1.36) that the value of A^0 and A^- characterize the relative of the charged and neutral forms of adsorption particles of the class abundance on the surface and could be considered as the probability of finding the particle adsorbed in some charged state. On this basis, the correlation between these values is such that

$$A^0 + A^- = 1. \quad (1.37)$$

If the level rises above the Fermi energy level of $2\ kT$, in the gap created by the adsorbed particle, then A^- is almost equal to 1 (Figure 1.5, a and b).

Lowering of the Fermi energy level below *2 kT* created by adsorbed particles, leads to a value of A^- is 0. If $E_2 = E_f$, then $A^- = 0.5$.

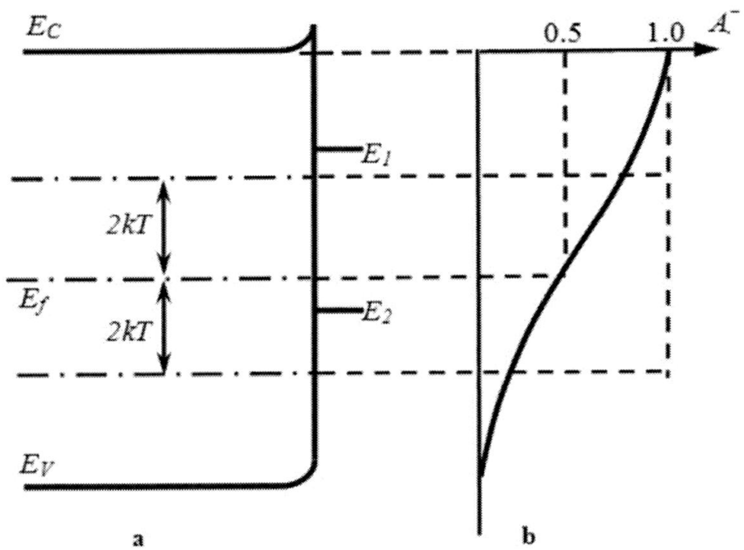

Figure 1.5. The band bending at the surface of semiconductor in the case of chemisorption. (see, please, explanations in the text).

Thus, it follows from the above, that the relative content of a form of chemisorption of the particles on a solid surface, is determined by the position of the Fermi level in the band gap at the surface of a solid. Adsorption capacity of the surface is the equilibrium value of the relative surface coverage for adsorbed particles.

Let the cross-sectional area of one of its own atom solid surface - adsorption sites be $s = 10^{-19}$ m^2. Consider the surface area $S = 1\ m^2$. Then the total number of adsorption centers that are located on the surface of $1\ m^2$:

$S/s = 10^{19}$, and the concentration of $N^* = 10^{19}\ m^{-2}$. If the total concentration of adsorption centers on the surface, containing the adsorbed particles on themselves is N, then the adsorption capacity of the surface is:

$$\beta = N / N^*. \qquad (1.38)$$

Thus, the determination of the adsorption capacity of the surface is reduced to the problem of finding β depending on the position of the Fermi level in the band gap of the solid.

Langmuir equation for the general case (1.3) for the steady state, gives the steady number during chemisorption of acceptor particles on a solid surface with n-type conductivity, so one could obtain:

$$0 = aP(N^*-N)-\beta N^-$$

$$aP(N^*-N) = \beta N^-$$

$$\beta = \beta_0 \exp\left[(E^- - E_f)/kT\right]$$

(β_0 is constant). According to the Weitz limit $N^- << N^*$. Then:

$$\alpha PN^* = \beta_0 N^- exp \left[\left(E^- - E_f\right)/kT\right]$$

where the adsorption capacity – is determined as:

$$\beta = \frac{N^-}{N^*} = \frac{\alpha P}{\beta_0}\exp[-(E^- - E_f)/kT] \tag{1.39}$$

Since $E^- < 0$ and $E_f < 0$ and $|E^-| > |E_f|$, then $E^- - E_f < 0$, and $\{-(E^- - E_f\}> 0$. Therefore, as the distance E_f from the bottom of conductivity band value $\{-(E^- - E_f\}> 0$ decreases, and therefore decreases the adsorption capacity of the solid surface $\beta = N/N^*$ for acceptor particles.

Thus, removing the Fermi level of the conduction band adsorption capacity of solid surfaces to acceptor particles decreases, and when approaching the Fermi level to the bottom of the conduction band it increases, i.e. the Fermi level acts as a regulator of the adsorption capacity.

1.7. Effect of Crystal Size in Its Adsorption Properties

In one of the possible cases of the dependence of the adsorption capacity of the size of the crystals start to appear, if the number of possible localization of carriers on the surface is comparable to the number of free electrons in the conduction band of the solid. Consider a crystal of $1\ cm^3$, which has an area equal to all the surfaces of *1* cm^2. Let every single surface atom plane *1* cm^2, with the limit of Weitz, 10^{13} localized electrons and the crystal has a high electronic conductivity - such that the concentration of electrons in the conduction band of 10^{17} cm^{-3}. By hypothesis, the number of electrons in a volume $V = a{\cdot}b{\cdot}l$ should be equal to the number of electrons on the surface area of $S = a{\cdot}b$, i.e.:

$$10^{17}\text{cm}^{-3}a{\cdot}b{\cdot}l = 10^{13}\text{cm}^{-2}a{\cdot}b. \tag{1.40}$$

where l is equal to 10^{-4} *cm*. Therefore, the influence of the surface in this case can occur only for sufficiently small crystals, which are:

$$V/S < l. \tag{1.41}$$

The value of l is actually the depth to which is bound the surface effect on the properties of the crystal. In many cases, its value is comparable to the Debye screening length.

Experimental examinations of the effects on the surface properties of the volume begin to appear at the thickness of crystals (10^{-4}-10^{-5} *cm*). In these crystals, begins to demonstrate itself the dependence of the position of the Fermi level of the ratio V/S [4, 6]. As we consider the case of the original model with *n*-type conductivity thickness $2L$ ($L >> l$), on both surfaces of which, are chemisorbed the particles retained on the electrons captured from the bulk of the plate (Figure 1.6) [1]. This leads to the vanishing of the band bending at the surface, while the volume of the plate remains neutral and has flat area (horizontal band position). Therefore, in this case, a fairly thick plate is considered and its volume remains insensitive to the processes on the surface. When the thickness of the plate, as compared to the depth distribution of influence on the properties of the surface of the is such that $L \leq l$, so the curvature of the zone boundaries of the allowed bands in the volume of the crystal, but to varying degrees (Figure 1.6, b). The greatest deviation of the zone boundaries from the horizontal position occurs at the surface and gradually decreases towards the middle of the plate, which also did not show the formation of flat zone boundaries. Fermi level is shifted on the surface and in the bulk of the plate in comparison with as well Figure 1.6, a.

Further reduction of the thickness of the plate can lead to a complete flattening of the zones on the surface and in the interior (Figure 1.6, c). In this case, the Fermi level position relative to the bottom of the conduction band at the surface and in the bulk are the same.

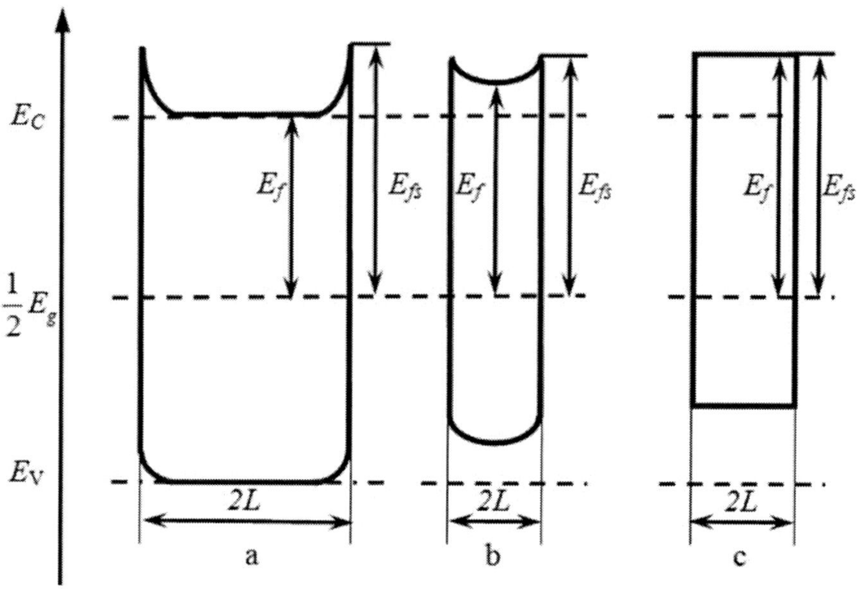

Figure 1.6. Change the position of the Fermi level with decreasing thickness of the plate. (see, please, the explanations in the text).

In Figure 1.6, the change in the position of the Fermi level is due to the result of fragmentation of the sample (or reduction of the size of the plate). This procedure leads to an increase in the distance between the Fermi level and the bottom of the conduction band, which, in turn, causes a decrease in the concentration of free electrons which could participate in the adsorption process and be captured by acceptor particles on the surface. Thus, reduction of the geometrical dimensions of the sample, finally, leads to a decrease in the adsorption capacity of the surface for adsorption of acceptor particles and, increases the adsorption capacity of the adsorbed particles of the donor type.

1.8. Alternation of the Characteristics of the Surface by Adsorbed Particles

1.8.1. The Surface Charge during Adsorption

As even on an atomically clean surface, there are initial energy levels of Tamm and Shockley types, the electrons are captured from the region, which causes the accumulation of negative charge on the surface [7, 8]. Equal to this, taken in absolute value, the positive charge is located on the ionized donors in the surface region of the crystal. This leads to an emergence of the negatively charged surface and the formation of a double layer, with the positive charge located near the surface of the solid.

Appearance of the adsorbed particles on the surface leads to the formation of the band gap on the surface of the local energy levels which have charge exchange with the volume of the solid [6].

There are several types of adsorption interaction of adsorbed particles with the surface [4]. As a result, some of them, after the formation of local energy levels of the adsorption origin, on these levels are trapped electrons or holes from the bulk of the solid, and the surface acquires the additional charge compared to the biographical defects, which are there even on the atomically clean surface.

If the particles are adsorbed on the surface with certain concentration of acceptor-type N^{-}, then there are the localized electrons taken from the surface region of a solid, and the appropriate forms of the negative charge density on the surface are in total number of:

$$\Sigma^{-} = eN^{-}$$

Adsorption of the donor particles on the surface with a concentration of N^{+}, which give their electrons to the conduction band and the positively charged surface emerges, the solid surface acquires an additional positive charge in a number:

$$\Sigma^{+} = eN^{+}$$

Equal to it, as to the absolute value of the negative charge, the charge is located near the surface of the crystal.

The total charge can be expressed as the sum of the negative and positive charges accumulated on the surface:

$$\Sigma = \Sigma^{-} + \Sigma^{+}$$

The magnitude and sign of the net charge depends on the ratio of the partial charges present on the surface. If the values of the negative and positive surface charges are exactly equal, the total surface charge is about zero. The surface should be neutral, despite the fact that the magnitude of the partial charges could be substantial. In this case, a flat energy zone area, and the Fermi level is located about the middle of the band gap.

If the solid surface of *n*-type, the Fermi level is above the middle of the forbidden zone, the surface dominates electron capture levels acceptor adsorption of origin and the total value of the surface charge is negative. The position of the Fermi level in the band gap at the surface below the middle of the band gap leads to the fact that the surface is positively charged.

On the neutral surface is no band bending, i.e., the boundary of the allowed energy bands are flat. The accumulation of negative charge on the surface energy levels leads to the formation of the locking band bending (up), that is, to the impoverishment of the surface region of the major carriers. If as a result of adsorption of the adsorbed particles transfer their electrons into the conduction band of the surface region, then it is here that the layer enriched with electrons, and the zones are bent downward.

Adsorption of foreign particles contributes to a solid surface and generates an additional charge of additional surface energy levels involved in the charge exchange with the volume of a solid.

1.8.2. Changes in the Surface Potential during Adsorption

Suppose that on the original neutral surface adsorption of acceptor particles is only one type at a concentration N_s, creating a band gap at the surface of the electron capture levels below the Fermi level, which are localized electrons from the conduction band. Surface charge eN_s causes value of the surface potential V_s. In considering the adsorption of donor particles, be aware that the sign of the surface potential will be the opposite [1, 4, 6].

In general, the equilibrium value of the surface potential, the steady-state after adsorption can be represented as the sum of the initial value of the surface potential (before adsorption) V_{so} and the change of the surface potential due to adsorption ΔV_s:

$$V_s = V_{so} + \Delta V_s$$

For a neutral surface $V_{so} = 0$. Original (historical) value of the surface potential is difficult to measure. Therefore experimentally in many cases using the Kelvin method (vibrating electrode) to measure changes in the magnitude of the surface potential of which is calculated on the charged surface states formed by the adsorbed particles. Concentration of ownership (biographical) energy levels on a solid surface, however, remains unknown.

1.8.3. Effect of Adsorption on the Work Function of the Solid

Important characteristics for changes which can be seen on the processes occurring on the surface, are thermionic φ_T and photoelectric φ_f work function

$$\varphi_T = E_f + V_s + V_d + \delta$$

$$\varphi_f = E_g + V_d + \delta$$

(E_f - the position of the Fermi level in the gap in the amount, E_g - band gap, V_s - surface potential barrier, V_d - dipole component of the work function, δ - electron affinity of the solid). The band gap, the position of the Fermi level in the volume and the electron affinity of the solid surface conditions do not depend on (Figure 1.7) and by the nature and condition of the material. The surface potential depends on the surface, because its value is expressed in terms of the concentration of charged surface levels. Dipole component of the work function due to the potential drop in the electric double layer created by the charges of which the adsorbed particles in the surface layer of the solid.

The value of V_d may be due to physically adsorbed particles, if they have a dipole moment, or particles that acquire a dipole moment as a result of adsorption on a solid surface.

Changes caused by adsorption of the thermionic $\Delta\varphi_T$ and photoelectric $\Delta\varphi_f$ work functions can be represented as follows:

$$\Delta\varphi_T = \Delta V_s + \Delta V_d$$

$$\Delta\varphi_f = \Delta V_d$$

The magnitude of changes in the thermionic work function can be determined by measuring the potential difference between the Reference electrode and the surface of the sample, if the pre-known changes in the work function of measuring indicator electrode in the adsorption of the same kind of particles. Some of the electrodes, for example, gold, such changes are negligible and can be ignored, which greatly simplifies the experiment.

Change of the photoelectric work function can be measured from the shift of the red edge of the photoelectric effect. Further comparisons of the results of measurements of changes photoelectron and thermoelectric work functions on the same sample and under identical conditions allows, by subtracting from $\Delta\varphi_T$ the value $\Delta\varphi_f$ of first determine the value of ΔV_s, and then from the expression $\Delta\varphi_T$ to find the value of ΔV_d. In most cases,

$$\Delta V_d << \Delta V_s$$

and changes of the surface potential can be seen directly from the measurement of changes of the thermionic work function, because in this case

$$\Delta\varphi_T = \Delta V_s$$

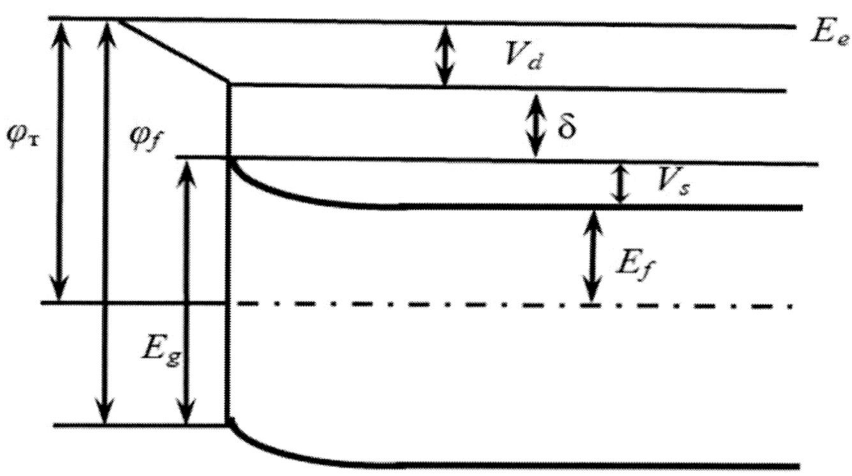

Figure 1.7. Band diagram for the determination of the thermionic and photoelectric work function: E_e - the position of the energy level of vacuum. The remaining notes in the text.

1.8.4. Effect of Adsorption on the Conductivity

In the adsorption of acceptor particles on a solid surface formed local energy levels, which in time are filled with electrons from the conduction band of the material [4]. If this process occurs at the surface of the material n-type conductivity, the concentration of free electrons in the sample is reduced and, consequently, reduced its conductivity:

$$\Delta\Sigma = \Sigma_1 - \Sigma_0$$

(Σ_0 - on the value of the electrical conductivity to the adsorption, Σ_1 - value of electrical conductivity after the adsorption of acceptor particles, in this case for the *n*-type semiconductor $\Delta\Sigma < 0$). If the surface of adsorbed donor particles, the concentration of free electrons in the conduction band increases and $\Delta\Sigma > 0$. It follows that the adsorption of acceptor particles on a solid surface with electron and hole conductivity change of electric conductivity of these materials have opposite signs:

n-type conductivity, $\Delta\Sigma < 0$;
p-type conductivity, $\Delta\Sigma > 0$.

Thus, knowing the type of acceptor particles, on the base of electroconductivity sign change measurements one can set the type of electrical conductivity of the material, without the Hall effect measurements. On the other hand, if the type of solid state conductivity (*n*-type or *p*-type) is known, on the base of the sign changes $\Delta\Sigma$ it is possible to determine which particles are adsorbed - acceptor or donor type.

1.9. Local and Collective Interaction on Semiconductor Surfaces

Contact semiconductor surface with the surrounding gas atmosphere, usually accompanied by a change in its electrical properties [9-18]. The size and speed of these changes depend on the electronic properties of the surface and volume of the semiconductor, the nature and reactivity of adsorbed particles [10-29].

Falling on the surface, the particles of the gas medium or leave some time back into the gas phase, or remain on the surface, held by the interaction with the semiconductor. These forces may be of different nature. If they are by Van-der-Waals forces of electrostatic polarization, electrostatic images or events caused by the dispersion, the particles are considered to be in a physically adsorbed state [30]. If communication with the surface of the particles of the gas formed by the overlap of the wave functions of the electrons of the adsorbate and adsorbent, and the interaction between them in the formation of chemical bonds, then the chemisorption [31-37].

For a description of both the chemical and physical properties of the surface of the most widely used two models. One of them is the atomistic model, or the model of local interactions. Another model - the band one, it is called also the model of collective interactions [38].

The atomic model is preferable when considering chemical processes on solid surfaces, where the most important role played by local interactions between atoms and molecules on the surface. In atomistic modeling of surface phenomena are described in terms of surface centers, i.e., atoms or groups of atoms belonging to the surface, and are considered primarily chemical interactions between them. The band structure of a solid state is almost completely ignored here. When atomistic approach considers only local connection alien (adsorbed) atom with the selected atom or group of atoms on the surface of the crystal and is virtually ignored the existence of a solid as a solid crystalline system. Naturally, this model has advantages when considering microscopic chemical interactions.

Model of collective interaction is more useful in the analysis of the processes of electron exchange between the volume of solids and surface based on the band model. In the model of collective interaction at the description of the surface one applies the concept of surface states, which correspond with the local surface electronic energy levels in the band gap and almost completely ignore the presence of microscopic interactions among the atoms of the surface.

For a long time a special attention was paid to the band model. In this model the electron is non-localized Bloch wave functions. The appearance of the surface levels associated with the fact that the adsorbed or other foreign atom introduces local defects in the periodic structure of the surface. As a result, these atoms form the surface states on the energy levels that can capture electrons or with which electrons can be excited into the conduction band of the solid. Any local (chemical) interactions among foreign atoms and the surface were not considered. This created the benefits of the band model, since in the framework of its mathematical description of the process is independent of the chemical nature of the adsorbate and details of local chemical interaction.

In 1947 year F.F. Wolkenstein, deciding the quantum mechanical energy spectrum of the crystal surface in the presence of her outside of the molecule, has shown that the chemisorbed particle can be considered as a structural defect in the surface, the ability to localize itself around a free electron from the conduction band or hole of the valence band of the semiconductor [39]. Similarly, the structural defect of the crystal defects created chemisorbed particle can be in neutral or charged. Neutral state of the defect is called a weak form of chemisorption. Charged - a solid form of chemisorption [40]. Position in the band gap of the semiconductor energy levels corresponding to the type of chemisorbed particles and their concentration, introduced phenomenologically as mean of payment in most cases it is not possible to accurately solve the problem of the quantum gas - semiconductor. Filling of electrons or holes of local energy levels created in the gap on the surface chemisorbed particles, and thus, they share the captured charge determined by the position of the Fermi level in the band gap of the semiconductor.

According to [41-43], the capture of electrical charges on the surface level results in the formation of the space-charge region (SCR) at the surface and is accompanied by changes in the electrical properties of the sample, such as electrical conductivity [44-47], the concentration [48-50], of mobility [51 -53] of the majority carriers, the lifetime of minority carriers, the electron work function of the semiconductor [54], etc.

The advantages of a team approach are in possibility to apply adsorption theories and research methods, widespread in semiconductor physics. Ease of interpretation of the chemisorption of problems with this position is achieved through a simplified representation of the interaction of the reactants with the semiconductor, which amount to the acquisition or loss of an electron. Thus, all the chemical properties of the chemisorbed molecules are reduced to the ionization potential and electron affinity, and all the chemical properties of the crystal to the position of the Fermi level, which in this context has a role as a regulator of not only the current transfer, but the chemisorption properties of a semiconductor. In particular, the establishment of electronic and chemisorption equilibrium relationship between different forms of adsorption is determined by the position of the Fermi level [1].

The development of the electron theory of chemisorption, based on an idealized model of a solid surface, allowed to predict the presence of natural connections between electrical and adsorption properties of semiconductors. On the other hand, the experiment is not always possible to establish a direct correlation between the consequences of the theory, the main parameter in the theory - Fermi level and the sensitivity of the adsorption surface of the semiconductor.

The main causes of discrepancies between theory and experiment lie in the fact that the theory considers an idealized surface, and the experimental results reflect the real properties of the surface. These discrepancies are also related to the fact that the main parameter of the theory - Fermi level - in the real world depends on many factors (temperature, pressure, concentration of biographical and extrinsic defects, etc.), some of which can not be reliably monitor and determine the current methods.

Electron theory of chemisorption indicates the fact of a local level, however, does not allow for the numerical calculations of their parameters. Therefore, research continues to address these problems. The achievements in the theory of chemisorption [1-48] in most cases, not yet provided the ability to predict the parameters of surface states, the adsorption properties of the surface and their changes during chemisorption. This is due to the fact that the solution of the problem is extremely difficult to account for all possible types of surface irregularities, the role of the nature of adsorption sites and adsorbed particles, etc.

Summary of surface states comes from experimental studies performed by Auger spectroscopy, low-energy electron diffraction, ultraviolet and *X-ray* photoelectron spectroscopy, infrared spectroscopy, *EPR*, etc. However, the application of these methods is quite limited, as their use is possible in the case of atomic clean surfaces, or difficulties related to the change of surface parameters under high excitation energies.

Consideration of the results reported in the literature shows that the use of modern methods of research has not led to a full disclosure of the relationship of local and collective mechanisms for adsorption-desorption processes, as a qualitatively new experimental techniques, along with studies of individual properties of adsorption complexes not provide for registration, for example, the concentration of and carrier mobility, which reflect the collective properties of the electronic system of the semiconductor. Therefore, the challenge remains associated with the development of new methods for the analysis and characterization of the real surface and adsorption levels.

With the advent of advanced mathematical methods and computer modern quantum theory in some cases, have managed to combine both approaches. In this analysis, a surface atom with a group of neighboring atoms is analyzed without neglecting the band structure of a solid. The results of these calculations in many cases are close to the experimental data.

Atomistic approach [49], using the model of frontier orbitals, considering the interaction of the energy levels of the relevant fragments of molecules using the perturbation theory, various "jelly"-substrates, methods of self-consistent pseudopotential semiempirical tight-binding method augmented plane wave method of calculation using the cluster approach [50, 51] and have been widely used field theory [52], and combinations of atomic orbitals (LCAO) [53].

Quantum mechanical analysis methods applied mainly to the ideal or, in the best case, atomically clean surfaces of solids and only in certain very specific conditions allow us to estimate the energy of the real surface adsorption states. However, the MO LCAO methods allow you to design the electronic levels of the cluster to a complex set of atomic wave functions, which is particularly useful when considering the adsorption processes in the framework of the atomic orbitals.

The development of quantum chemistry and establishing simple rules stereochemistry showed that molecular orbital theory can be applied effectively by analyzing the interaction of molecular fragments in the contour diagrams. Sketches of the molecular orbitals more clearly show that the displacement of electrons occurs only if the orbitals came into contact and overlap. Combining sketches all occupied molecular orbitals gives a picture of the overall distribution of the electron density in many-electron molecules. A graphical representation of the molecular orbitals is useful for elucidating the mechanisms of the interaction of molecules with the surface of the solid.

Obviously, the boundary orbitals of the chemisorbed molecules interact in different ways with different surface centers, selecting those orbitals, which are topological structure similar to the structure of the orbitals of the adsorbate. An important role in determining the local characteristics of sorption subsystem and, therefore, a factor that controls the passage of the reaction, is also a type of hybridization of the orbitals of the adsorbate molecules, determines the degree of excitation and independent of external conditions.

An important consequence of the application of the model to the local interaction of the chemisorbed molecule with the solid is that the energy level position, appearing in the band gap of the semiconductor during chemisorption, along with the nature of the chemisorbed molecule is largely determined by the energy position and the configuration of the frontier orbitals of surface defects, which interact the molecule. Therefore the determination of the nature of the main center of chemisorption under appropriate environmental conditions is a major challenge in clarifying the mechanism of chemisorption.

Combination of the collective and local approaches in frame of the model that explores elements of system analysis makes possible to use it for description of chemisorption processes and their influence on the electrical conductivity in the surface-active structures with varying concentrations of the surface of the charged defects.

In the atomic model, the term "surface facility" refers to a group of microscopic atoms on the surface (Tab. 1.1), which in a sense is active. Centers may belong or be a uniform surface in heterogeneous areas of the surface, where their activity is often higher. Superficial atom of the host lattice with a broken (free) bond, bond-free bonding orbital with the large value of the electron affinity or occupied bonding orbital characterized by a low ionization potential could be viewed as the superficial centers on the uniform clean surface. Center on the surface of an ionic crystal can be one of the cation, which is not offset by a corresponding number of anions, as is the case in the bulk, and detects high electron affinity. The density of the centers of this type reaches 10^{19} m^{-2}, i.e., one center for each atom of the surface. On the other hand, centers may be associated with defects of non-uniform surface. Examples of such centers can be centers for crystallographic steps or at the intersection of dislocations with the surface. In the case centers on the steps of the crystallographic foreign atom (adsorbed particles) can interact with several atoms of the lattice, the resulting total interaction in this case is very strong.

Admixture on the surface can also act as a hub, as well as a mixture of different oxides. In the band model of localized surface states are responsible for the surface electronic energy levels. These levels tend to actively exchange or share electrons with non-localized energy bands in the bulk of a solid (conduction band and the valence band).

Table 1.1.

The atomic model, or model of local interactions	Band model or a model of collective interactions
1. Analyzed the chemical processes in which implemented local interactions between atoms and molecules on the surface.	1. The processes of electron exchange between the surface and the amount described in framework of the band model of a solid.
2. Surface processes described in terms of surface centers. Ignored band structure solid.	2. Uses the concept of surface states, which correspond to the surface energy levels. Microscopic interactions are neglected.
3. Surface core-atom on the surface with dangling bonds is a free bonding orbital.	3. The appearance of the surface energy levels of surface states - is the result of violations of the periodic structure of the surface or the result of her appearance on a foreign atom.

Therefore, when considering the surface states are important their macroscopic properties, not the microscopic atomic orbitals, which are responsible for the existence of surface states [35, 38].

For the case of an atomically clean surface in quantum mechanics is proved that the localized energy levels on a solid surface can occur simply because of the fact that the surface gives the periodicity of the crystal lattice. These surface states are often referred to their own state, or the state curriculum. The causes of these states may be different. For example, the state of Shockley occurs mainly on the surface of covalent substances to the extent that the orbitals of the valence electrons of atoms strongly overlap. The associated overlap exchange energy stabilizes configuration valence electrons in the solid.

On the other hand, the surface atoms of some of the electrons are in orbitals that are perpendicular to the surface (dangling bonds), and do not overlap with neighboring orbitals. Levels that differ from the levels of the valence electrons in the bulk correspond to these electrons.

Another example are the states arising in cases where the surface atom is different than an atom in the bulk, the electron affinity, that is, a violation of the periodicity of the potential move. The electrons of the surface atoms are at levels that do not match the energy of the energy levels in the bulk. Thus, in the case of a band model with a cadmium ion present on the surface of the solid as an impurity may change its charge state to Cd^{2+} due to the transition of an electron in the conduction band. This process proceeds according to the equation:

$$Cd^{+} \rightarrow Cd^{2+} + e.$$

At the zone scheme both ions Cd^{+} and Cd^{2+} corresponds to one and the same energy level, only changes its charge state. If the ion of another substance, for example, has a greater ability to give electron, its energy level is located on the zone scheme is closer to the bottom of the conduction band. Consequently, between the chemical properties of the impurity atoms, the adsorbate and the expected energy level of the surface state he established a direct link.

On the other hand, in the strict model of surface centers (atomistic model) interaction with the allowed energy states in the areas of solid-state is ignored, and the fore the local interaction with neighboring atoms. As part of this the interaction of the atom with a solid ion Cd^{+} is very different from its interaction with Cd^{2+} ions, so that these two cases correspond to two very different local interactions. Theory of chemical valency is applied for description of these interactions.

As already noted, on the surface may be different surface states. The differences between them are determined by the difference between their nature and origin. All possible types of surface states can be summarized and divided into three groups (Table 1.2).

Table 1.2. Three main groups of the surface states

1	Surface states of foreign particles on the surface.
2	Ionic surface states with unpaired electrons on the surface of ionic crystals - Tamm surface states.
3	Shockley surface states on the surface of covalent crystals with broken free bonds, which are responsible for the surface for covalent bonding.

The first group includes surface states created on the surface of a solid foreign particles and impurities adsorbed substances. These states capture or donate electrons in their interaction with the solid [35, 38].

The second group consists of ionic surface states, unpaired electrons of which can come or go away from the allowed energy bands of solid or from adsorbed atoms or molecules. If the ion is an acceptor surface state, it is associated with the cation, if a donor - that is associated with the anion. In addition to the term "ionic surface state" to describe these states use the name "Tamm surface" state, to distinguish them from the states Shockley typical covalent solids.

The third group - the surface states formed by dangling bonds on the surface of covalent solids. They appear in the form of unpaired electrons on the surface atoms with an incomplete coordination sphere. This type of the surface states can exchange electrons with the allowed energy bands or form adsorption sites with strong covalent bonding.

References

[1] Волькенштейн, ФФ. Физико-химия поверхности полупроводников. М.: Наука; 1973.

[2] Langmuir, I. Evaporation, Condensation and reflection of molecules and mechanism of adsorption. American. *Phys. Soc. Rev.*, 1916 vol. 8, 2217–2245.

[3] Рогинский, СЗ. *Адсорбция на неоднородных поверхностях*. М.: Изд-во АН СССР; 1948.

[4] Волькенштейн, ФФ. Электронные процессы на поверхности полупроводников при хемосорбции. М*.: Наука;* 1987.

[5] Lennard-Jones, JE. Processes of adsorption and diffusion on solid surfaces. *Trans. Faraday Soc.,* 1932 vol. 28, 333–359.

[6] Smyntyna V. *Electron and Molecular Phenomena on the Surface of Semiconductors*. New York: NOVA Publishers; 2013.

[7] Tamm, Jg. Uber linemogliche Act der Elektronenbindung an Kristalloberflachen. Zeitschrift fur Physik. 1932 vol. 76, Issue 11–12, 849–850;

[8] Shokley, W. On the Surface States Associated with a Periodic Potential. *Phys. Rew*., 1939 vol. 56, 317–323.

[9] Новотоцкий-Власов, ЮФ. Исследование влияния адсорбции на спектр электронных состояний реальной поверхности Ge, Si, антимонида индия: автореф. дис. д-ра физ.-мат. наук. М.: 1976.

[10] Зарифьянц, ЮА. Взаимодействие молекулярной и электронной подсистем на поверхности полупроводников и диэлектриков: автореф. дис. д-ра физ.-мат. наук. М.:1976.

[11] Литовченко, ВГ. Управление электрическими характеристиками с помощью адсорбции молекул. *Полупроводниковая техника и микроэлектроника,* 1972 вып. 9, 92–126.

[12] Козарицкий, ВД; Козлов, СН; Киселев, ВФ; . Новотоцкий-Власов ЮФ. Влияние анион- и катионрадикальных форм адсорбции на электрофизические параметры реальной поверхности германия. Доклады АН СССР, 1970 т.195 no.1, 115–118.

[13] Малинова, ГВ. Исследование хемосорбции атомов кислорода на ZnO методом КРП. *Журнал физической химии,* 1971.т. 46 вып. 1, 131–133.

[14] Быкова, ТТ. Влияние адсорбции кислорода и окиси углерода на величину потенциала поверхности CdS. Вестник Ленинградского университета, 1976 no. 22, 57–60.

[15] Shapira, Y. Chemisorption, photodesorption and conductivity measurements on ZnO surfaces. *Surface Science*, 1976 vol.54, no. 1, 43–59.

[16] Lagowski, J; Sproles, ES; Gatos, HC. Quantitive Study of the charge Transfer in chemisorption oxygen chemisorption on ZnO. *Journal of Applied Physics,* 1977 vol. 48, no. 8, 3566–3575.

[17] Göppel, W. Reactions of oxygen with ZnO— 1010 surfaces. *J. Vac. Sci. and Technol.*, 1978 vol.15, no. 4, 1298–1310.

[18] Briones, F; Golmay, D; Ortis, C. The role of oxygen in the sensitization of photoconductive PbSe films. *Thin Solid Films*, 1981 vol. 78, no. 4, 385–395.

[19] Зарифьянц, ЮА;. Карягин, СН; Киселев, ВФ. О возможности управления формами связи адсорбированных молекул путем изменения электронного состояния поверхности полупроводников. Доклады АН СССР, 1972 т. 202, №.1, 1402–1412.

[20] Bhide, VG; Jatar, S; Pastogi, AS. Kinetics of oxygen chemisorption and its effects on electrical conductivity of thin evaporated CdS films. *Pramana J. Phys.,* 1977 vol. 9, no. 4, 399–410.

[21] Bube, RH. The basic significans of oxygen chemisorption on the photoelectronic properties of CdS and CdSe. *J. Eletrochemical Society*, 1966 vol. 113, no. 6, 793–798.

[22] Каламазов, РУ. Роль серных вакансий в радиационной хемосорбции кислорода на сульфиде кадмия. В: Радиационные дефекты в полупроводниках. Минск : Изд-во БГУ; 1972.

[23] Malinova, GV; Myasnikov, LA. The effect of chemisorptions of atomic oxygen on conductivities 0f semiconducting oxides; 1.Pyrolysis and photolysis of oxygen. *Kinetics and Catalysis*, 1969 vol.10, 263-268.

[24] Нымм, УХ. Влияние поверхностного барьера на скорость хемосорбции на CdS. *Физика и техника полупроводников*, 1974 т. 8, вып.11, 2111–2115.

[25] Татабадзе, ДГ; Мясников, АИ. Взаимодействие атомов Н и О с тонкими пленками элементарных полупроводников. *Журнал физической химии*, 1975 т. 49, вып. 2, 367–370.

[26] Кировская, ПА; Зелева, ГЛ. О взаимодействии водорода и кислорода на поверхности алмазоподобных полупроводников. Журнал физической химии, 1978 т. 52, 1744–1747.

[27] Розова, ТТ. Адсорбция кислорода из германия, стимулированная сильным электрическим полем. Физика твердого тела,1978 т. 30, 2675–2674.

[28] Kutra, J; Zindulis, A; Sakalas, A. Influence of thermal treatment in various media on electric properties of polycrystalline CdSe layers. *Lithuanien Journal of Physics*, 1980 vol.20, no. 3, 110-113.

[29] Gogoi, S; Barua, K. Electrical conduction and breakdown phenomenon in dry air stabilized CdS films. Japan. *J. Appl. Phys.*, 1979 vol. 18, no.12, .2233–2236.

[30] Petitjean, IP; Moortzat, G. *Adsorption physique des gaz par les solides*. Silicat. ind., 1975 vol.40, no. 9, 221–230.

[31] Clark, A. *Chemosorptive Bond. Basis Concepts*. New-York— London: Academic Press; 1974.

[32] Lyo, SK. Theory of chemisorptions. In: *Topies in Applied Physics Interactions on Metal Surface*. New-York: Springer; 1975 vol. 4, 41–62.

[33] Crimley, TB. Chemisorption Theory. In: *Progress in Surface and Membrane Science*. 1975 vol.9, 71–161.

[34] Wolkenstein, Th. Randschichttheoric und Electronentheoric der Adsorption an Halbleitern. *Z. Phys. Chem*., 1977vol. 108, 97–104.

[35] Somorjai, GA. Surface Science. *Science,* 1978 vol. 201, no. 4355, 489–497.

[36] Adamson, AW. *Physical chemistry of surfaces*. 3 ed. New-York: Wiley; 1976.

[37] Englerl, L; Lorenz, W., Chemisorption on semiconductor surfaces generalized expression of partial charge injection and adsorption energy. *Surface Science*, 1981 vol. 104, no. 2–3, 549–558.

[38] Morrison, SR. *The chemical physics of surface*. New-York: Plenum Press; 1977

[39] Волькенштейн, ФФ. Электронные уровни атомов, адсорбированных на поверхности кристалла. *Журнал физической химии*, 1947 т. 21, № 7, 317–321.

[40] Волькенштейн ФФ. Физико- химия поверхности полупроводников. М.: Наука; 1973.

[41] Gatos, HC; Lagovsky, J., Space change layers Surface Eff. Cryst. Plast.In: *Proc. NATO Adv. Study Inst*. Mohegeiss. 1975.Leiden: 1977, 221–265.

[42] Сугата, Т. Адсорбция и электропроводность в металлических пленках и полупроводниках. Джикацу, 1973 т. 18, 1, 24–34. (пер. ВПЦ № Ц 19603).

[43] Frankl, F. Surface space charge layers. In: *Solid State Surface Science*. New York: 1973, vol. 3, 105–163.

[44] Ржанов, А.В. *Электронные процессы на поверхности полупроводников*. М.: Наука; 1971.

[45] Ляшенко, ВИ. Редактор. Электронные процессы на поверхности полупроводников. К.: *Наукова думка;* 1968.

[46] McRay, EG; Hagstrum, HD. Surface structure. *Ecsperimental Methods Treatise on solid state chemistry*, 1976 vol. 6 A, 57–61.

[47] Гринев, ВИ; Киселев, ВФ. Изменение электрофизических свойств реальной поверхности германия при адсорбции атомов водорода. *Вестник* МГУ. Сер.: физика, астрономия, 1981 т.22, № 2, 79–82.

[48] Doyen, G. Structural and Dynamical Aspects of Adsorption and Desorption. *Top. Curr. Phys.*, 1987 vol. 43,. 301–346.

[49] Che, M; Bond, GC. *Adsorption and Catalysis on Oxide*. Amsterdam: Science Publishers B. V; 1985.

[50] Karlson, CJ; Landmark, E; Johansson, LSO. [and others] Hydrogen chemisorption on Si (111) 7X7 studied with surface-sensitive cor-level spectroscopy and angl-resolved photoemission. *Physical Review B*, 1990 vol. 41, no 3, 1521–1528.

[51] Tamaki, J; Nagaishi, M; Teraoka, Y. [and others] Adsorption behavior of CO and interfering gases on SnO_2. *Surf. Sci.*, 1989 vol. 221, 183–196.

[52] Rappe, AK; Goddard, WAHI Charge equilibration for molecular dynamics simulations. *J. Phys. Chem.,* 1991 vol. 95, 3358–3363.

[53] Rantala, TT; Rantala, TS; Lantto, V. Surface relaxation of the (110) face of rutile SnO_2. *Surf. Sci.,* 1999 vol. 420, 103–109.

[54] Голованов, ВВ. Характеризация полупроводниковых поверхностно активных веществ $Sn_xW_yO_z$, $CdxS(Se)_{1-x}$, In_2O_3. : дис. д-ра физ.-мат. наук: 01.04.10 научный консультант проф. В.А.Смынтына; ОГУ им. И. И. Мечникова.Одесса: 1998.

Chapter 2

PHENOMENOLOGICAL ASPECTS OF ADSORPTION SENSITIVITY OF SEMICONDUCTOR GAS SENSORS

ABSTRACT

This chapter examines the definition of adsorption sensitivity, introduced by the author and co-workers. Criteria and limits of applicability of the definition of the sensitivity of the adsorption of materials are considered. Its advantages in respect to other methods of determining the adsorption sensitivity are analyzed. On the basis of this definition of the adsorption sensitivity adsorptive properties of crystalline, polycrystalline, macro-, micro-, nano-porous and amorphous semiconductors are considered.

2.1. ADSORPTION SENSITIVITY OF SEMICONDUCTOR RESISTIVE ADSORPTION-SENSITIVE SENSOR

The interaction of the semiconductor film *n*-type acceptor with gases reduces the conductivity of the sample due to the capture of conduction electrons of chemisorbed particles, such as atoms or molecules of oxygen. Due to the chemisorption of oxygen, the conductivity of films of cadmium selenide, as well as other materials, is significantly reduced. And, after the chemisorption reduces its conductivity, the reference value can be restored, due to the heating in vacuum. Therefore, it was natural to assume that the adsorption-desorption sensitivity (ADS) is the relative change of the layer conductivity [1]

$$\gamma = \frac{\sigma - \sigma_h}{\sigma_h} \tag{2.1}$$

where σ and σ_h - the electrical conductivity in the and adsorbtion- and desorbtion-influenced states.

Despite the fact that the definition of ADS, is given by (2.1), the considered property of the sensor itself along with its shortcomings are obvious. As follows from Figure 2.1 (curve 1), in the concentration range $\Delta C_1 = C_3 - C_2$ and $\Delta C_2 = C_3 - C_1$ value γ will have approximately the same value (0.85 and 0.90 arb.units), although taken in concentration of $C_1 - C_2$ is practically absent in the sample's ADS. Error in the γ value will grow with increasing concentration measurement range.

ADS definition as a measure *m*, depending on [2]:

$$\sigma = A \cdot p^m \tag{2.2.}$$

(A - coefficient, p - pressure) in most cases is not applicable when the measurement conditions (temperature, pressure range, etc.) are the variables.

Technical standards GOST 20220 for ADS, specifies the change of the signal at the output of the measuring circuit due to the change of the gas concentration at about 1 %. It is obvious that in this case, the value of ADS will depend essentially on the parameters of the circuit elements (amplifiers, bridges, input and output impedance, etc.).

From the analysis of the existing methods for ADS determination, it comes out that none of them makes it possible to compare the results of each ADS measurement, obtained not only on different materials, but also on the sensor of the same material, if it does not use at least a strictly identical intervals p, T and the same measuring circuit. This, along with the lack of analytical studies ADS relationship with biographical defects and properties of the semiconductor, respectively, adsorption methods for predicting the sensitivity, significantly complicate the development and application of semiconductor ASE.

Thus, it is evident that the definition of the parameters ADS semiconductor sensor. As such features can be used, for example, the electrical conductivity or resistance of the semiconductor.

Consider ADS as the relative change of electrical conductivity of the film due to the change of the gas concentration on 1 % vol., or by changing the partial pressure *p*, respectively [3, 4, 5]

$$\beta = \frac{\partial\Omega}{\Omega\partial C} \rightarrow \frac{\partial\Omega}{\Omega\partial p} \tag{2.3}$$

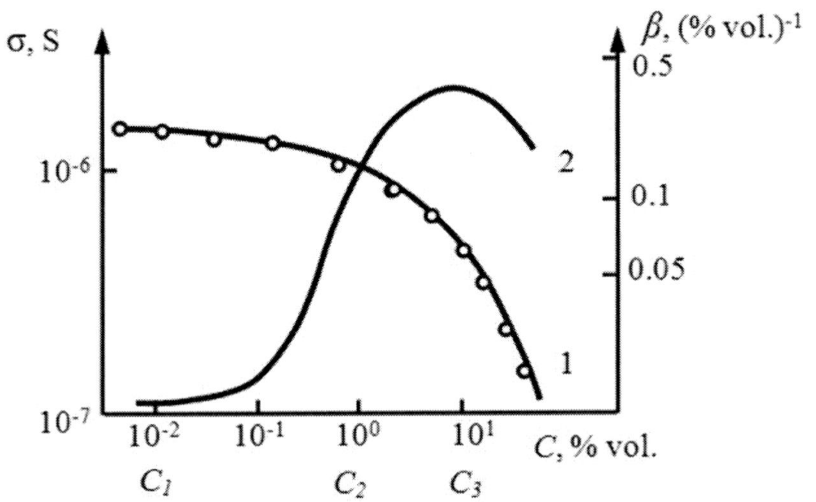

Figure 2.1. The dependence of the electrical conductivity (1) and the adsorption sensitivity (2) of *CdSe* film on the oxygen concentration.

From (2.3), it follows that the dependence of *β(c)* or *β(p)* could be obtained through numerical differentiation of the curve *Ω(C)* or *Ω(p)*. Here *Ω* – the conductivity set of the semiconductor layer at a concentration of gas *C* (or its partial pressure *p*).

As an example, Figure 2.1(curve 2) shows the result of treatment with the aid of *Eq.* (2.3) (curve 1), the same figure. In contrast to the data, obtained by other methods of determining the ADS, from the curve 2 the maximum value *β* is reached in concentration range (10^0-10^1) (% vol.)$^{-1}$.

Obviously, in the case of linear dependence $\sigma = f(C)$ to determine the *β*, is fair

$$\beta \approx \frac{\Delta\Omega}{\Omega\Delta P} \tag{2.4}$$

The use of (2.3) allows us not only to significantly improve the accuracy of measurements β, but also enables to correctly match each other in the parameter β various sensors, regardless of the measurement conditions.

Increase in the concentration of charge trapped chemisorbed particles changes the surface charge Q_s and the dimensionless surface potential Y_s, taking into account that the change of Q_s and Y_s are the causes of changes in surface conductivity and conductance G_s throughout the film [6], the expression (2.3) in the form [3, 4, 5]

$$\beta = \frac{1}{\Omega}\frac{\partial\Omega}{\partial Y_s}\frac{\partial Y_s}{\partial p} \tag{2.5}$$

Using the effects of space-charge region theory and the known expression for the surface conductivity [3, 4, 5, 7], we obtain the dependence of β on semiconductor parameters and external conditions

$$\beta = \frac{L_D\left[\lambda(\exp(-Y_s)-1+b\lambda^{-1}(\exp Y_s-1)\right]}{2(bd\lambda^{-1}+\frac{1}{2}L_D\int_{Y_s}^{0}\frac{\left[\lambda\exp(-Y_s)-1+b\lambda^{-1}(\exp Y_s-1)\right]dY}{F(Y_s\lambda)})F(Y_s,\lambda)}\times$$

$$\times\frac{A}{P_0(\exp(E_{td}-Y_s)\left[n_iL_D(P(Y_s\lambda)-F(Y_s\lambda)\right]-N_{dn}f_1(2-f_1)}. \tag{2.6}$$

A new approach to the definition of ADS allowed to obtain (2.6), the relation between β and major bulk and surface properties of the semiconductor and the experimental conditions [3, 4, 5-7].

Numerical calculations on a computer using the formula (2.6) are satisfied [3, 8] in the following characteristics of the system $CdSe$ - O_2 values of variables: $\lambda = 10^{-4}$, $d = 10^{-6}m$, $E_{ta} = 0{,}75\ eV$, $E_{td} = 0{,}3\ eV$, $N_{dn} = 10^{18}\ m^{-2}$, $\mu_p = 10^{-4}m^2/V\cdot c$, $b = 10$, $E_g = 1.72\ eV$, $L_D = 10^{-7}\ m$, $A = 10^{19}\ m^{-2}$, $P_o = 1\ atm$, $T = 500\ K$, — 30 *units* $\kappa T < Y_s < 20$ *units* κT. It was assumed that the oxygen chemisorbed in the form O_2^-, and the dissociation of molecules is missing.

Analysis of the results shows [4, 5, 8], that the most significantly β depends on the surface potential Y_s. For the considered semiconductor film of *n*-type *CdSe*, ADS has a maximum value at low enrichment (5 *units* $kT < Y_s < 10$ *units* kT) for surface electrons (Figure 2.2) and the inversion of the conductivity ($Y_s \approx -27$ *units* kT).

In case of deep depletion (-20 *units kT* $<Y_s<$-5 *units kT*), the main contribution to Ω the volume makes σ_{sc} component of the electrical and low temperatures β be virtually absent. As we approach the Y_s to zero, increases the contribution of G_s, which causes a sharp rise in β, which lasts as long as the positive charge of the surface donor biographical origin becomes more negative charge localized on the surface of chemisorbed oxygen molecules. When this condition on the curve $\beta = f(Y_s)$ has a maximum, in the future (Y_s > 10 *units kT*) β decreases, because change ΔG_s by chemisorption are much less than the value of G_s.

In the area of -30 *units kT* $<Y_s<$-20 *units kT* observed inversion of the sign of the majority carriers. Increasing the number of negative oxygen ions on the surface in this area Y_s increases the concentration of free holes p and surface conductivity G_s, and hence, β. The decrease of β under the condition that $Y_s<$-25 *units kT* is caused by the depletion of the oxygen level E_{ta} in the result of Fermi level approaching to it.

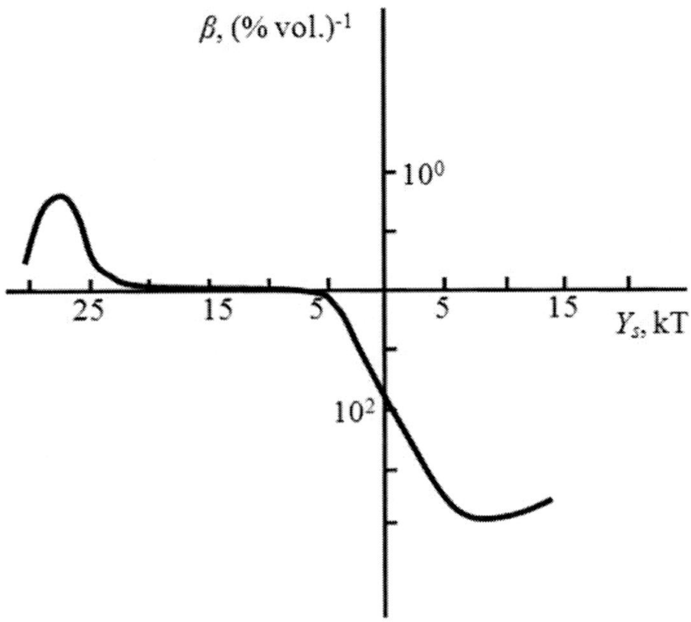

Figure 2.2. Change of the β as a function of surface potential. The calculation is performed using the formula (2.6), the values of the parameters used are given in the text.

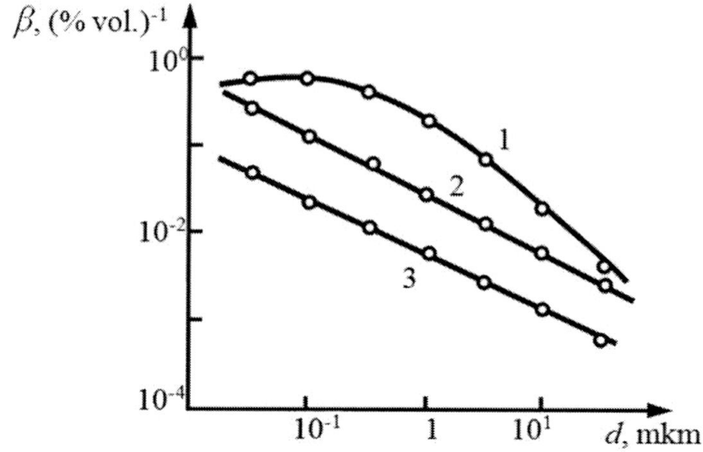

Figure 2.3. Changes in adsorption sensitivity β for sensor (*CdSe* film) with increasing thickness. The calculation is performed using the formula (2.6) for $Y_s = 7$ *units. kT* (1); - 4 *units. kT* (2); - 15 *units. kT* (3).

Reduction in the degree of deviation from the *n*-type semiconductor its own $(n_i / n_{io} = \lambda \to 1)$, the growth of the donor concentration at the surface, increasing the binding energy of the adsorbed molecules with the surface, leads to the increase in ADS. This is due to the influence of the bulk component of the decrease of conductivity σ_{sc}, straightening zones on the surface and, as well, in an increase in the filling of electron levels of adsorption origin, respectively.

The sensitivity β depends on the single geometric parameter - the thickness of the semiconductor layer, which is caused by the dependence of the bulk conductivity on *d*.

As in the case of enrichment (Figure 2.3, curve 1) and depletion (curves 2, 3), the surface β decreases with increasing *d* due to an increase in the contribution of the bulk component Ω.

Decrease in β, with increasing temperature is the result of a significant increase in the concentration of conduction electrons, leading to an increase σ_{sc}, and increasing the difference $(P - F)$ in the denominator of (2.6). With further increase in *T*, dominated by β *(T)*, λ has increased, which leads to higher ADS (Figure 2.4).

If the concentration of N_{dn}, comparable in magnitude to the maximum possible concentration of chemisorbed ions, in an area where $N_{ta} << N_{dn}$ (area of low concentrations of O_2), β is small. When the conditions of deep depletion are reached, i.e., in the area - 20 *units kT* < Y_s < 5 *units kT* (large concentrations of *C*), β also is virtually absent. In the intermediate range with dependent β *(C)* passes through a maximum, which position depends on the value of N_d.

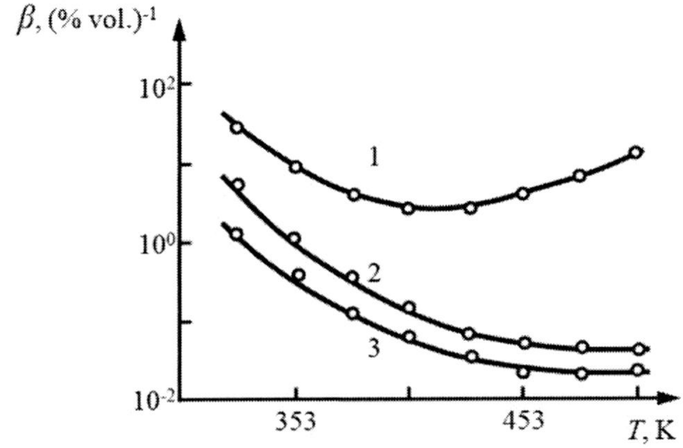

Figure 2.4. The temperature effect on the sensitivity of the sensor β for *CdSe* film. The calculation is performed using the formula (2.6).

If the concentration of N_{dn} is so large that for any value *C* fair $N_{dn} > N_{ta}$, then the growth will occur only an increase in β. If there are biographical depletion on the surface with increasing *C* (N_{ta}), then will decrease β. Thus, in the experimental study of the dependence $\beta(C)$ three types of curves can be found, described by only decreasing or increasing function of only $\beta(C)$, or containing a maximum at a certain value of *C*.

2.2. Adsorption Sensitivity for Crystalline and Polycrystalline Adsorption-Sensitive Semiconductor Sensor

If the equation of (2.6) is not taken into account as described in [5], the uneven distribution of the properties will be found through the thickness of the semiconductor film, such as cadmium selenide. As major donors both be centers of adsorption, the increase in the vacancy concentration of selenium in the initial stages of growth can lead to increased β, as well as the "leakage" of cadmium atoms on the surface due to the increase N_{dn}. If in this case we consider that during film growth increases the contribution to the overall conductivity of G_s, it is clear that the experimentally observed increase in β with increasing d (Figure 2.5) can not be fully described by (2.6) and is due as described in [5] process self-ligation film cadmium selenide. It also shows that after the process begins to develop self-compensation self-ligation that ultimately leads to the crystallization of the film and, consequently, to a decrease in N_{dn}. After a sharp decrease G_s main factor influencing the β, is inversely proportional relationship σ_{sc}^0 of d, and ADS decreases. Thus, only at sufficiently large d, where the cadmium selenide film is recrystallized in a stable monophase system, we should expect the correspondence between the calculated and experimental dependences $\beta(d)$, that is indeed the case (see Figure 2.3 and 2.5).

The obtained experimental dependence $\beta(T)$ at low temperatures (Figure 2.6) does not match the calculated data. Only at $T > 450$ K, the predicted reduction in 2.1 β corresponds with increasing T.

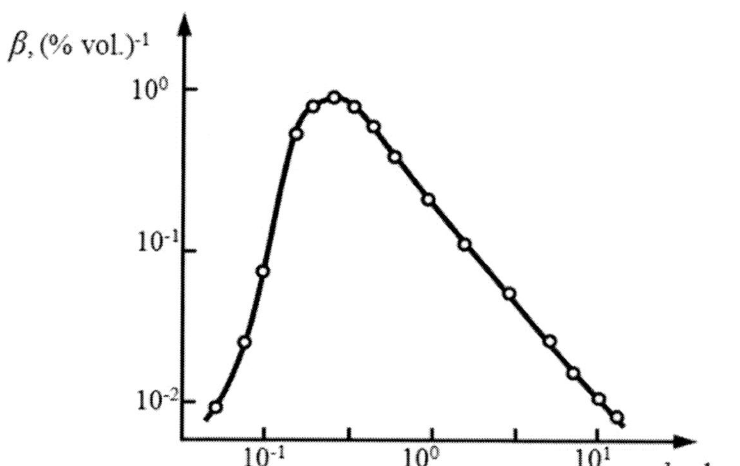

Figure 2.5. Experimental dependence of the adsorption sensitivity β on the thickness d of the film of cadmium selenide. Measuring temperature is 420 K.

Studied potential difference and the kinetics of electron-molecular processes have shown [5] that the increase in the bed temperature stimulates the transition from molecular to atomic form of oxygen chemisorption. What is happening with the sharp increase in the E_{ta} (from 0.75 *eV* to 1.2 *eV*) and N_s (from $5{\cdot}10^{18}$ to $9{\cdot}10^{18}$ m^{-2}), by (2.6), should lead to an increase in β.

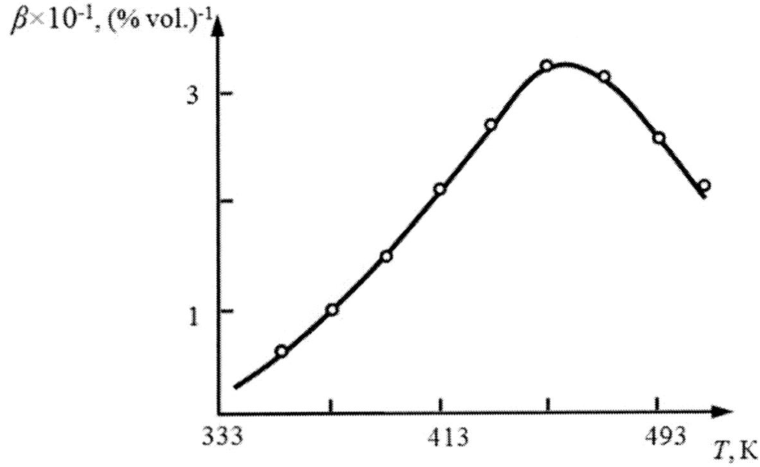

Figure 2.6. The dependence of the adsorption on the temperature sensitivity of the β film cadmium selenide, measured experimentally.

Characteristically, the samples of differing nature of the major donors occur in the surface region, as the maximum on the curve $\beta(E)$ is obtained at different temperatures. The maximum value of the samples' β, is caused by major donors which occupy the selenium vacancies, observed at 410 *K*. For films with major donors corresponding to interstitial cadmium atoms, ADS begins to decrease at $T > 460$ *K*. These films differ in the basic form of chemisorbed oxygen on the surface [5]. In the first case, the main form of chemisorption is atomic oxygen, in the second - O_2. Consequently, the position of the curve $\beta(T)$ depends on the composition of the semiconductor surface, which determines the relationship between the forms of chemisorbed oxygen.

2.3. Porous Semiconductor Sensor's Adsorption Sensitivity

Unlike resistive ASE with relatively dense structure, porous semiconductors have a much more highly developed surface. In this case, under the influence of surface, adsorption processes can significantly change and contribute to the ADS not only in the resistance along the surface, and the resistance in the direction perpendicular to the surface, as well as the change in the capacitance of structures that make up the sensor. Consequently, for the porous structure is necessary to analyze change in the total impedance of the adsorption-sensitive element. Convenient to perform such analysis on silicon, in which the pores are already technologically obtained macro-, micro-and nano-scale [4].

For macroporous material with transparent contacts for gas gap type (Figure 2.7, a), the circuit diagram is shown in Figure 2.7, b. Plot circuit between terminals A, B (Figure 2.7, b) in the absence of rectification measurements can be replaced by an equivalent resistance $R = (R_2^{-1} + (2R_1 + 2R_3)^{-1})^{-1}$. Here R_1 and R_2 – sample resistance along the surface and perpendicular to the interior of the material, R_3 - the contact resistance of the interface porous-*Si* - *C: Si*. Alternating current measurements in sandwich mode between terminals A and *C* equivalent capacitance of the structure is approximately equal to C_2, and the planar direction between terminals A and B $C = C_2/2$. Impedance structure is $Z = (\omega^2 C^2 + R^{-2})^{-0.5}$. The adsorption sensitivity of impedance *Z* macro-porous silicon semiconductor structure can be written as:

$$\beta_Z = -Z^{-2}\left\{\omega^2 C^2 \beta_c - R^{-2}\beta_R\right\}, \tag{2.7}$$

$\beta_C = C^{-1}dC/dc$, $\beta_R = R^{-1}dR/dc$ — values of the structure capacity and resistance. The value of the crystalline *Si* R_3 is sufficiently smaller than the resistance of the porous *Si*, that's why the input of R_3 is negligible. Then, as $R = (1/R_1 + 1/2R_2)^{-1}$, ADS will be equal to:

$$\beta_R = R^{-1}dR/dc = (1/R_1 + 1/2\,R_2)^{-1}\{R_1^{-1}\beta_{R1} + (2R_2)^{-1}\beta_{R2} \tag{2.8}$$

Then, it is rational to analyze the ADS separately, considering the values of sensitivity β_{R_1}, β_{R_2} and β_C.

During the consideration of the adsorption sensitivity β_{R_1}, the resistance R_1 is the resistance over the surface of the porous material. Considering the idealized scheme of macro-Por-*Si*, it is convenient to consider that the pores are distributed uniformly over the surface between the contacts (Figure 2.8).

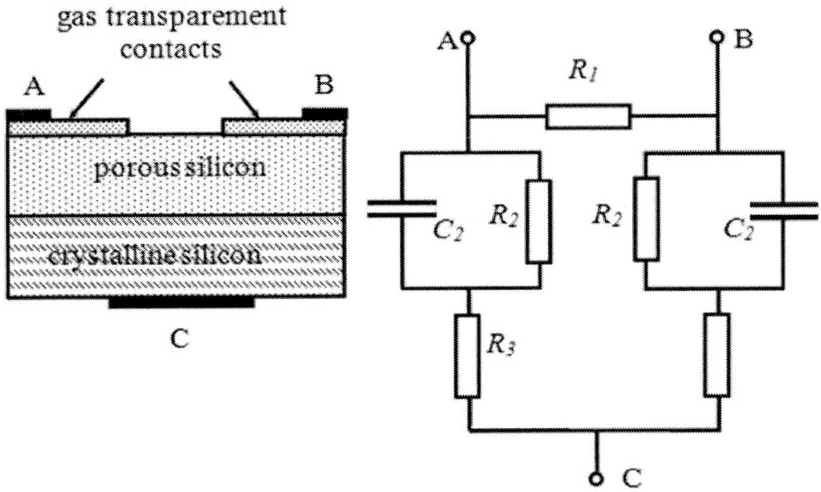

Figure 2.7. The structure of the sample of porous silicon formed on a crystalline silicon thin film with electrical transparent aluminum contacts for gas to porous silicon (A, B) and the electrical contact of the aluminum to the crystalline silicon (C) (a) and simplified equivalent circuit diagram of a sample of porous silicon (b). R_1 - resistance along the surface of the sensor, R_2 - resistance perpendicular to the surface, R_3 - the contact resistance of the interface.

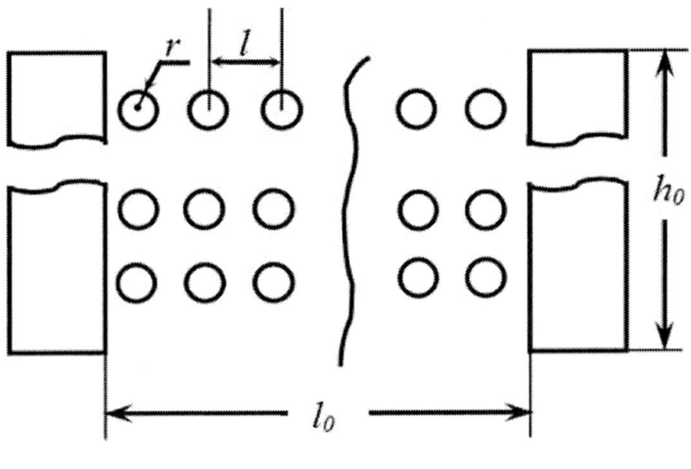

Figure 2.8. The simplified scheme of the macro-porous *Si* (r — pore radius, l — distance between the pores' centers, h_0, l_0— width and length of the inter-electrode space).

Let it be dominated as follows: r – pore's radius, l — distance between the pores' centers, then the resistance R_1 of the periodic structure Por-*Si* represents the sum of the resistances R_{ij} and R^*_{ij} in the form of the grid. For the contacts used being of A and B types, the sensitive layer has the geometric values: *d*— porous Si layer thickness, l_0 — the length of the inter-electrode distance, h_0 — the width of the contact. In this case, between the electrodes are over the pores $m = {l_0}/{l}$ and along the width of them over $n = {h_0}/{l}$ the pores [4].

Under the constant voltage between the anode and cathode, the knots where the resistances are connected R^*_{ij}, are situated on the equipotent lines. Thus, in accordance with Kirchhoff laws, the current does not flow between these lines, and the equivalent scheme is represented by series and parallel resistances R_{ij}.

$$R_{Ij} = \sum_j R_{Ij} = nR_0;$$

$$R_1 = \left\{ \sum_i \left(nR_0\right)^{-1} \right\}^{-1} = nR_0 / m \quad , \tag{2.9}$$

and equivalent resistance of the scheme is equal to R_0n/m (R_0 — being the resistance of one cell of the grid) and

$$\beta_{R_1} = R_1^{-1}\, {dR_1}/{dc}$$

$$ {dR_1}/{dc} = \left(R_0 n / m\right)^{-1} \times d\left(R_0 n / m\right) / dc = R_0^{-1} \times dR_0 / dc \tag{2.10}$$

Thus, it comes out of (2.9), that the ADS of one layer of macro-Por-*Si* depends on the ability of one cell to change its resistance under adsorption. The width and length of the inter-electrode space doesn't influence the ADS value.

The general resistance of one cell could be represented by parallel connection of two series resistances R_0^*. The cell resistance is equal to $R_0 = R^*$. Thus, the problem lies in the determination of the resistance R_0 dependence on the technological factors and to detect the ability of this resistance to change its value under the influence of gas adsorption.

If the radius of the *i-th* pore r_i is greater than 1/2, and the porous channels of the greater radius are rare and the silicon structure remains crystal, then at the adsorption of the donor gases, the *n*-conduction channel is realized and the adsorption sensitivity doesn't depend on the sample geometry, the volume properties of the porous silicon is determined by its surface properties at the other equal conditions.

The sensitivity value β_n for the donor gas is always positive for the silicon of the *n-type*. The size of the high electron conductivity increases at the electron concentration increase in the vicinity of the pore.

In the samples of *p-type*, in the case of depletion, the adsorption sensitivity (ADS) sign is negative, what is caused by the hole conductivity region's hollowing at the increase of the hole depletion in the region of the pore's surface. In the case of the small distances between the pores, the connection of the electron conduction in the silicon samples of the p-type of inversion. As the result the ADS sign becomes positive again.

The sign of the AD sensitivity β_μ is negative. It certifies that with the gas pressure increase, the carriers' mobility decreases. The exclusion of the parameters r close to *1* at the surface potential value close to zero is the special case. As the values β_n and β_μ are summed according to the definition β_{R_0} , then the ADS absolute value will depend on the values and signs of the components.

For the ADS increase, it is necessary that, the β_n and β_μ have the similar signs. This case is realized, when the surface is depleted or the inversion channel is created on the surface.

In other cases, the ADS will be determined by the difference of β_n and β_μ. The sign of ADS depends on the value greater by the module. Under the significant distances between the pores, the conductivity will dependent slightly on the adsorption processes.

Under the approach of the inter-pore distance parameter to zero in the samples with hydrogenized region under the surface, the high ADS increasing has to be registered. The greatest ADS value occurs at the surface potential values close to zero.

2.4. The Changes of the Adsorption Sensitivity at the Transfer from Macro- to Micro- and Nano-Porous Silicon

In general case, the adsorption sensitivity β_Z depends on the β_R and β_C (2.7). From the results and calculations, presented in [4], it follows that during the enlargement of the average porosity of the macro-porous *Si*, the parameters β_{R_1}, β_{R_2} and β_{R_3}, increase. This, in turn, causes the ADS increase β_Z. In the region of medium porosity (≤25 %), main input to β_Z is made by term with β_R. In the case of the high porosity *Si* samples (porosity > 56 %) at the frequencies > 10 *κHz*, the main input is caused, determined by β_C. The sign of the parameter β_C depends on the sign of the surface potential. If there is the depletion on the surface of *n-type* macro-Por-*Si*, then at the chemisorption of the donor type gas, the surface potential β_C will be negative. At the slight enrichment of the surface, the same process will lead to $\beta_C > 0$. That's why the conditions could be realized, presented in (2.7)

$$\omega^2 C^2 \beta_c - R^{-2} \beta_R = 0$$

and, consequently, the changes of impedance *Z* are absent, despite the adsorption occurs and the great enough changes β_R and β_C in other values.

Under the transfer from macro- Por-*Si* to the structures of micro- Por-*Si*, the e.m.f. in micro-porous *Si* is observed, being placed into the NH_3 atmosphere and illuminated by white light spectrum. Under the porosity increase from 12 % to 43 %, the e.m.f. value increased from (3–10) *mV* to 35 *mV*.

The e.m.f. creation under the influence of NH_3 atmosphere and illumination (Figure 2.9), as shown in [4], is connected with the adsorption of the polar molecules at the sites with potential minima which are situated asymmetrically over the surface aligned to the direction of the applied electric field.

Under the transfer to the nano-porous *Si*, the rare luminescence effect is registered with the maximum at E_L = 1.7 *eV* (Figure 2.10), which intensity with diminished NH_3 concentration adsorbed over the structure surface. The changes of luminescence intensity were registered in the time intervals about 1 sec after the NH_3 concentration changes.

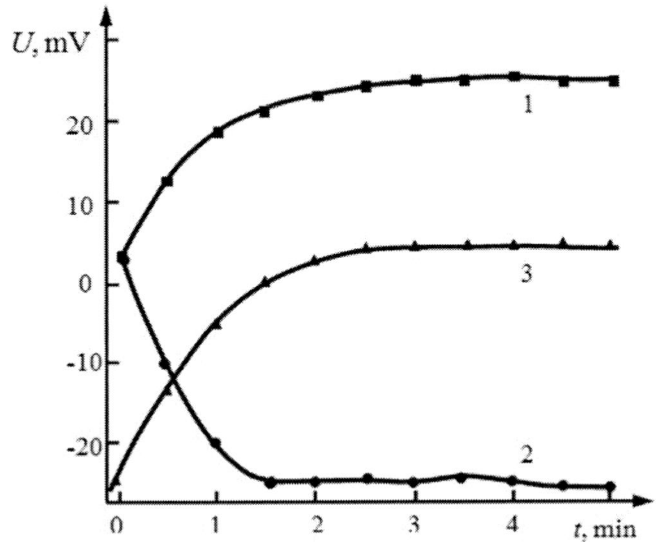

Figure 2.9. The kinetics of e.m.f. value changes *U* in micro Por-*Si* under the illumination with white light of 130 *lux* intensity in vacuum (1), and adsorption of NH_3 vapor of concentration 490 *ppm* in the dark (2) and under illumination of the said intensity after the NH_3 adsorption (3).

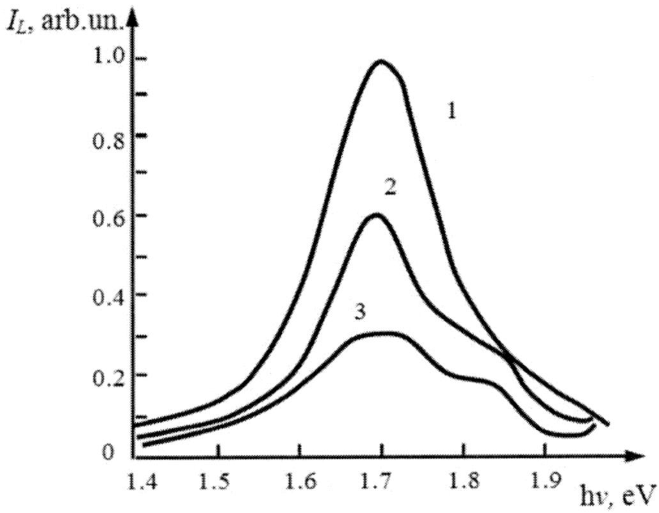

Figure 2.10. The spectra of nano-Por-Si photoluminescence, measured in pure N2 (1) and in mixture of NH3 of concentration 110 ppm (2) and 490 ppm (3) with pure N2.

This fact indicates that NH_3 molecules are placed at the surface of nano-Por-*Si* in the physically adsorbed state and are blocking the centers of radiative recombination placed over the sample's surface. The experimental procedures under study, allow to discriminate new phenomena and effects which surfacial accommodation is increased due to the transfer to the nano-Por-Si.

2.4.1. Adsorption Sensitivity β_{R2}

In the sandwich regime of measurement with gas-penetrative electrodes, the current flows into the depth of the layer through the resistance R_2. Considering the pores being distributed uniformly over the Por-*Si* surface, then adsorption sensitivity is determined by the properties of one cell [3, 4, 8, 9]. As well as in the previous case, the zones inclination leads to the changes of concentration and carriers' mobility values in the subsurface part of the pore which are determined by the formula

$$\beta_{R_2} = R_2^{-1} dR_2 / dc = (-G_2^{-1} dG_2 / dy_{so}) \times (dy_{so} / dc) \tag{2.11}$$

For the case of the stable crystal properties, the adsorption sensitivity is determined by the value dy_{so} / dc and doesn't depend on the geometry of the pore. The maximal value of this parameter is determined by the values of the surface potential, the binding energy and adsorbate's concentration on the surface.

The carriers' mobility in the channel is also determined by the surface zone inclination. The greatest mobility changes are registered directly at the surface of the pore.

2.4.2. Adsorption Sensitivity β_C

The samples of the porous silicon have the significant value of capacity, which changes by tens of times during the NH_3 vapor adsorption. The studies of the volt-capacitance characteristics have shown that the value of this capacity depends insignificantly on the voltage applied to the structure. This certifies that the physical origin of the samples' capacity is not connected with the presence of the diode structures, but with the presence of the significant surface and this dielectric layer in the capacitor. One of the capacitor's plates is formed by the crystal Si matrix and the second could be the free surface of Por-*Si*. During the NH_3 vapor adsorption, the great amount of energy states are formed on the pore surface. The NH_3 molecules' adsorption causes the increase of ionic conductivity along the *Si* surface [4, 9]. Thus, the second plate of the capacitor could be presented by the whole surface of the porous silicon. At the vapor pressure growth, the ionic conductivity along the surface and the greater surface of the sample is included into the process of capacity formation. This is a probable mechanism of the capacity growth due to the gas concentration increase [4].

Let us assume the uniform distribution of same size pores over the surface. Then, the each cell capacitance is included in parallel and the summary capacitance C is equal to:

$$C = \sum_i \sum_j (C_{lij} + C_{2ij}) = mn(C_1 + C_2)$$

Here m, n — number of pores over the length and width between the electrodes, C_1, C_2 — cell capacitance over the side surface and bottom of the pore. Then adsorption sensitivity will be equal to:

$$\beta_C = C^{-1}dC/dp = \left[mn(C_1 + C_2\right]^{-1} d\left[mn(C_1 + C_2\right]/dc = \\ = (C_1 + C_2)^{-1} d(C_1 + C_2)/dc.$$

Thus, the value of adsorption sensitivity is determined again by the properties of one cell, formed in the pore zone. Let us consider the values of the capacitances C_1, C_2 separately. The capacitor C_1 is connected with the pore's geometry: one plate has the pore's radius r_0, the second is equal to the half of the inter-pore distance $r = 1/2$. The capacitance C_2 is connected with the pore's bottom and the pore radius r_0. Thus, the values of these capacitances could be presented by:

$$C_1 = \varepsilon\varepsilon_0 d / \ln\left({}^{r_1}\!/_{r_0}\right),\ C_2 = \varepsilon\varepsilon_0 \pi r_0^2 / (r_1 - r_0).$$

The values of these capacitances are compiled by two parts: the capacitance of the space charge region and the least part of the capacitance. The value of the space charge region capacitance, as a rule, is much greater, because the charges of different sign are situated at the minimal distance in it. These two capacitances are included in series.

As a result of the adsorption processes progress, the main changes of the capacitance are due to the changes of the space charge capacitance region. That's why, the adsorption sensitivity could be determined through the derivative C_{SCR} over Y_s.

References

[1] Рыльцев, ИВ; Гутман, ЭЕ; Мясников, ИА. Влияние "эффекта поля" на адсорбцию атомов и молекул водорода и кислорода на окиси цинка. *Журнал физической химии*, 1981 т. 55, 968–976.

[2] Котелков, ВН. Влияние газовой среды на электропроводность ряда окисных полупроводников и некоторые возможности использования. В: *Физика полупроводников и полупроводниковых приборов*. Саратов: Изд-во Саратов. ун-та;1968 вып. 1, 171–174.

[3] Вашпанов, ЮА;. Смынтына, ВА; Сердюк, ВВ. Чувствительность электропроводности полупроводниковых пленок к хемосорбционно-десорбционному взаимодействию с газовой средой. В: *Анализ газов с применением современных средств микроэлектроники и полупроводниковой техники. Сб. трудов. ВНИИАП*. К: 1982, 24–33.

[4] Вашпанов, ЮА; Смынтына, ВА. *Адсорбционная чувствительность полупроводников*. Одесса: Астропринт; 2005.

[5] Smyntyna, V. Electronic mechanism for adsorptive sensitivity in semiconductor gas sensors. *Sensors and Actuators B*, 1994 vol. 19, 426-428.

[6] Ржанов, АВ. *Электронные процессы на поверхности полупроводников*. М: Наука; 1971.

[7] Сухарев, ВЯ; Мясников, ИА. Об электропроводности неупорядоченной полупроводниковой системы с межкристаллитными барьерами при адсорбции активных частиц. *Физика твердого тела*.1985 т. 27, вып. 3, 705–709.

[8] Вашпанов, ЮА. Адсорбционная чувствительность полупроводниковых материалов A_2 B_6, оксидов тяжелых металлов и кремния с реальной поверхностью с кластерными структурами. *Дис. д-ра физ.-мат. наук*, научный руководитель Смынтына, ВА. Одесса: ОГУ; 1999.

[9] Смынтына, ВА., научный руководитель. Исследование электронно-молекулярных явлений на поверхности полупроводниковых слоев и разработка научных основ физико-химических процессов в полупроводниковых сенсорах.: *Отчет о НИР (заключительный.)*: Тема № 161, Одес. нац. ун-т им. И.И.Мечникова ; Одесса: 2002.

Chapter 3

ADSORPTION SENSITIVIY OF THE REAL SURFACE OF SEMICONDUCTOR GAS SENSORS

ABSTRACT

The results of studies of the surface composition and the nature of adsorption centers on the interaction of semiconductor surfaces with adsorbed particles are discussed in this chapter. The mechanisms of the changes in the surface composition, the nature of adsorption centers, adsorption sensitivity and the main form of chemisorption of particles that occur during the production of materials for gas sensors also are discussed. It is shown that there are physical reasons why the property of the surface can not be reproduced from sample to sample, taken from the same production process.

3.1. ADSORPTION SENSITIVITY OF THE REAL SURFACE OF SEMICONDUCTOR MATERIALS

The semiconductor and metal-oxide adsorption-sensitive elements (ASE) are obtained through the use of different technologies depending on the targets established and the properties of the materials under consideration. Among the technologies applied, one could mention: molecular beam in open vacuum [1], the film deposition in quasi-close volume [2], electro-hydrodynamic liquid dispersion [1, 3, 4], chemical deposition on the substrate from the solution [5], thick film technology and gravure-offset printing [6–9], magnetron dispersion [10] and other technologies.

The specifics of each mentioned above technologies lies in that the ASE obtained through their use should obtain:

1. the surface with high adsorption sensitivity;
2. the dependence of the signal obtained with the ASE use on the adsorption processes progress on the surface.

The first condition could be satisfied by great majority of the technologies due to various technological approaches, used in the production of the microelectronic devices.

The second condition puts the efficacy of the ASE in dependence of its thickness.

It's evident that, in the case of resistance ASE, the greatest changes in its resistivity over the whole volume will occur when its thickness will be comparable with Debye length. In this case, the electron-molecular processes occurring on the ASE surface will influence the concentration of the free current carriers as well as their mobility in the whole ASE volume. These conditions are achieved better in thin semiconductor and metal-oxide films. It was shown both experimentally and theoretically, that these films could be considered to be thin at $d \leq 1$ *mkm*, e.g. in the thicknesses' range, where the surface carriers' scattering could be significant and their mobility significantly depends on d [11, 12].

In thick ($d > 1$ *mkm*) films the influence of the surface scattering is negligibly small and μ becomes independent on d. The height of the inter-crystallite barriers is a significant factor both in thin film ($d < 1$ *mkm*), ASE and in thick film ($d > 1$ *mkm*) ASE, which determine their electric properties [11, 13].

It is possible to reach the high ASE sensitivity with the use of the special refinement of the surface of both thin- and thick-film ASE, while the thin high adsorption-sensitive film is created on the surface of the element. This point is discussed in details in Chapter 4.

According to the theory of chemisorption, the increase of the free electrons' concentration and that of the electro-conductivity of the n-type semiconductor should be accompanied by its adsorption sensitivity to the adsorption of the gas particles of acceptor type [14], e.g. oxygen atoms or molecules.

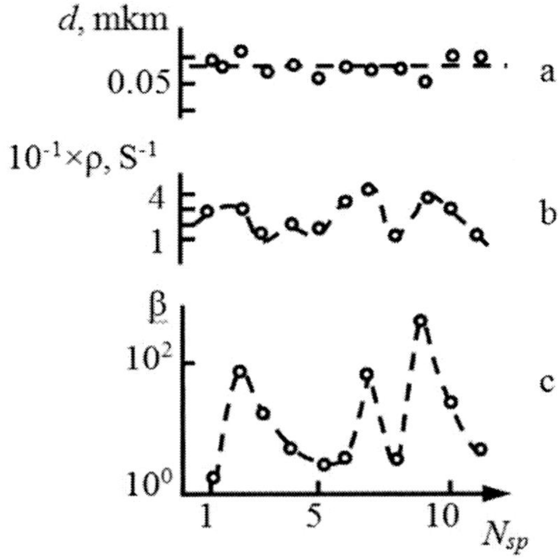

Figure 3.1. The dependence of the thickness (a), resistivity (b) and adsorption sensitivity (c) on the cadmium selenide sample number.

The electron theory of chemisorption is being based on the simplified assumptions of the solid state surface with the particles provided from the ambient.

These idealized conditions are not always fulfilled in reality and thus, in many cases, the results of even simple experiments are contrary to the theoretical prognosis [1, 15–18]. The verification of this thesis could be found in 2.5 of [1].

The sense of the registered inconsistency lies in the following. The *CdSe* films were obtained through molecular beam sputtering on the dielectric substrates obtained in the same technological cycle. Such films' thickness was equal in the range of the measurement error.

As to the ρ value, the samples differ by 2-3 times without any seen cases. It should be noted, as well, that according to the value of the adsorption sensitivity signified as β (Figure 3.1), the *CdSe* films could differ by more than 10^2 times.

And, in contradiction to the theory, the more resistive films could demonstrate β being greater than the value of the adsorption sensitivity of the less resistive samples.

The analysis of the CdSe samples' surface chemical content, studied in [1, 19–22], has shown that the discovered differences by the values β and ρ of the samples of equal thickness are caused by the differences of the main centers of chemisorption nature and main donors. It was assumed that the main donors on the real non-doped, in great majority of the cases, are, at the same time, the main adsorption centers [1, 20, 22]. This fact is clearly presented at the distribution of the *CdSe* films, which parameters are presented at Figure 3.1, into two groups according to the nature of the main donors (Figure 3.2). In such a case, the clear bonding between β and d as well as for the samples, for which the main donors are the selenium vacancies (E_d =0,14 eV), and for the films, the electro-conductivity of which is controlled by the inter-nodular atoms of cadmium (E_d = 0,3 eV) [1, 19].

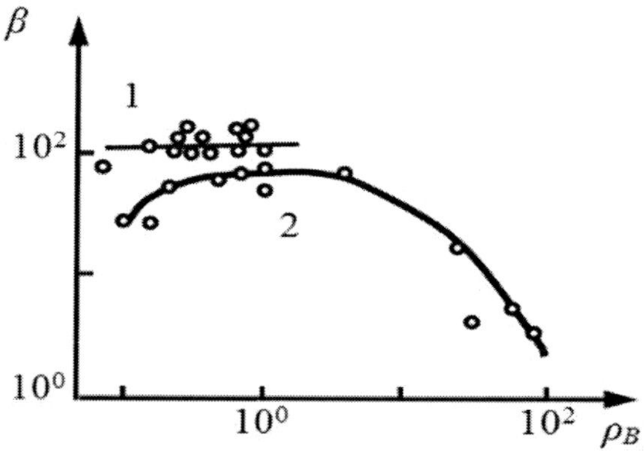

Figure 3.2. Dependence of the adsorption sensitivity β on the given resistance of the just sputtered *CdSe* films of the group 1 E_d = 0,14 eV (1) and of the group 2 E_d = 0,3 eV (2)

It was established, as well, that the selenium vacancies fulfill the role of the chemisorption centers for atomic oxygen, while the inter-nodular cadmium atoms serve mainly as chemisorption centers for oxygen molecules, e. g. the nature of the main chemisorption centers as well as the main donors in non-doped films determine not only the value of adsorption sensitivity but also the form of the chemisorption of the external particle [1, 19–27].

3.2. The Changes of Adsorption Sensitivity in Process of Semiconductor Films Growth

In the process of the films thickness growth, its resistance diminishes proportionally to the film thickness d [28–30], what presents the effect determined by the geometrical factor. If in the process of the *n*-type conductivity film thickness growth, the functional dependence $\rho = f(d)$ declines from the linear one, and diminishing, goes to the saturation, one could state that this fact indicates the defects' of *n*-type storage. During this process, the each following layer being created with *d* growth has the value ρ smaller, than resistance of the former layer [28–30].

The storage of the inter-nodular cadmium in the structure of *CdSe* film leads to the diminishing of the ρ along with the *d* growth (Figure 3.3). The increase of the internal defects concentration in the film structure causes the recrystallization into hexagonal modification. In process of over-crystallization, the super-balancing cadmium is being extracted on the film surface and is registered in a form of clusters. The size of *Cd* clusters could approach 1 *mkm* [31–35]. The diminishing of the internal donors' concentration as a result of the over-crystallization and extraction onto the surface, leads to the increase of ρ (Figure 3.3). Further, the film growth process along with the storage of the proper donor defect storage could go on over this surface accompanied with the diminishing of ρ (Figure 3.3).

The changes of the surface films' content, occurring during the process of their growth determines the dependence of adsorption sensitivity on *d* (Figure 3.4). It was registered, that the increase of cadmium atoms' concentration on the surface causes the adsorption sensitivity increase. The further super-layering of the *CdSe* film on the surface of the recrystallized substrate leads to the creation of the new layers containing less cadmium quantity on its surface and β along with *d* growth diminishes (Figure 3.4). Further, the process of film self-doping could be started again and, as a result, the adsorption sensitivity increases with *d* increase. It follows from Figure 3.3 and Figure 3.4, as well as from the papers [31–35], that in the process of semiconductor films growth, the occurring changes of their surface content determine the details of non-monotonous dependence of surface adsorption sensitivity on their thickness.

The value of the proper resistance (proper electro-conductivity) is the macroscopic parameter depicting the volume properties of the material while the adsorption sensitivity β reflects the properties of its surface determined by the interaction of external particles with the proper surface defects. That's why it is not possible to expect the simultaneity or equality of the changes of the volume and surface properties. But, as the main donors could serve as the main chemisorption centers, e.g. of oxygen on the surface of *CdS, CdSe* or other films, the inter-correlation between the changes of β and ρ could have place [31–35].

It is possible to strengthen the inter-connection between β and ρ (or σ), and in such a way, to increase the reliability of the information obtained through formation of the main channel of the current transfer over the surface or in the subsurface region of the film.

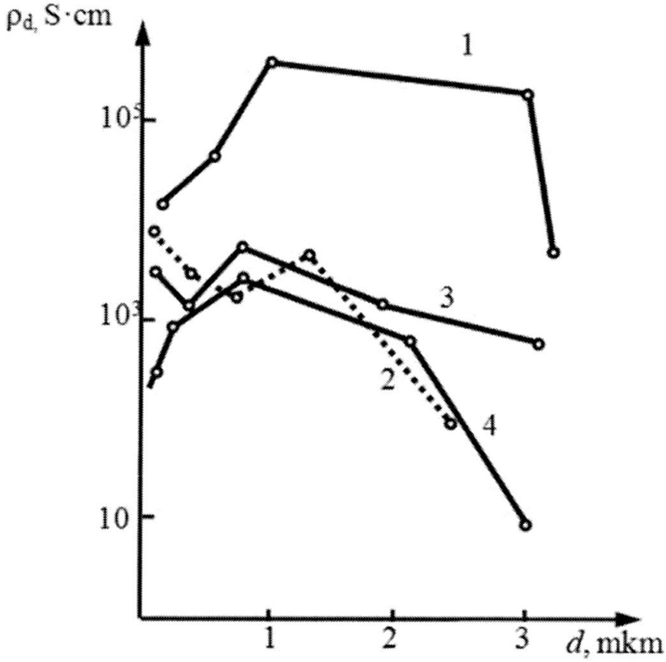

Figure 3.3. Dependence of the proper resistance ρ_d on the thickness *d* of the *CdSe* samples, obtained at the following substrate temperatures: 1) 150 °C, 2) 200 °C, 3) 250 °C, 4) 300 °C.

Such high sensitivity to the chemisorption of the acceptor-like gaseous particles, such as oxygen, could be obtained through termination of their creation at a certain stage when the surface is enriched with *Cd* atoms in the case of *CdS* and *CdSe* [31–35] films (Figure 3.3, Figure 3.4).

As it follows from Figure 3.1 and Figure 3.2, the reproducibility of the results will be low enough and such method can't be recommended for the sensors production.

But, these results demonstrate directly that the doping of the surface with the atoms being donors to the semiconductor of *n*-type conductivity, leads to the increase of adsorption sensitivity.

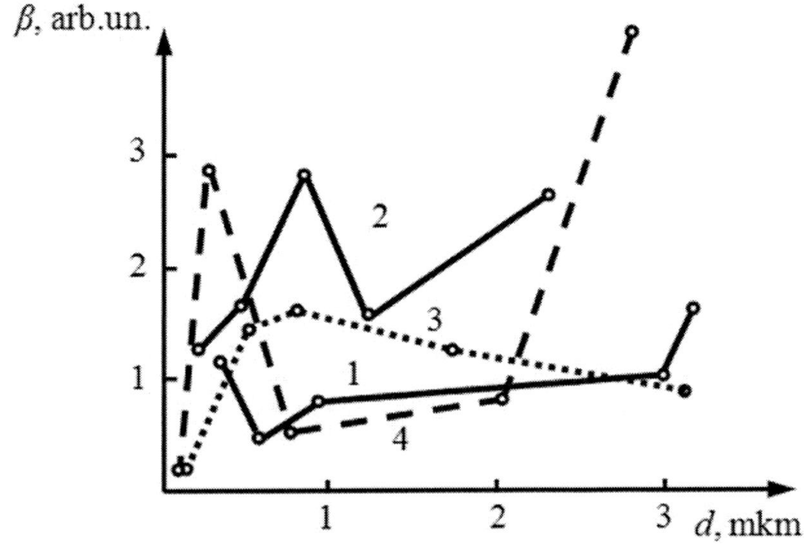

Figure 3.4. Dependence of the adsorption sensitivity β on the thickness of *CdSe* films, obtained at various temperatures of the substrate: (1) 150°C, 2) 200 °C, 3) 250 °C, 4) 300°C).

This is confirmed as well by the results of the surface potential's studies along with the thermoelectronic work-function of the *CdS* and *CdSe* films for the samples characterized by the creation of *Cd* clusters extracted from the volume to the surface during the film growth [1, 34]. For the starting *CdS* films, practically insensitive to O_2 chemisorption, the spectrum of the contact potential difference (CPD) contains only one peak thin enough at 4,16 *eV*, which is caused by the charge localized at the biographic surface states with the concentration $N_{s\delta} \approx 2\times10^{15}\,m^{-2}$. After surface self-doping with *Cd* atoms, the surface becomes sensitive to the chemisorption of the oxygen [1, 23, 33] and on the CPD spectrum the additional peak occurs at 4,7 eV, which is caused by the charge storage at the surface level of the adsorption nature with the concentration of $N_{sa} = 1{,}1\cdot10^{16} m^{-2}$ [1,33]. The chemisorption nature of this maximum is supported by the fact that it disappears at the samples' heating in high vacuum.

3.3. Adsorption Centers on the Surface of Semiconductor Films

In the basement of the effect of adsorption sensitivity increase, caused by the increase of the concentration of the proper donors on the surface, could be different causes. If the energy states of the proper donors are situated higher than Fermi level, so they could give their electrons into the conductivity zone and remain positively charged. This positive charge compensates the negative biographic surface and diminishes, in such a way, the surface potential barrier φ_s. In turn, the decrease of φ_s causes the increase of free electrons' concentration captured from c-zone on the energy states created by the chemisorbed improper atoms and molecules. At these conditions, the adsorption sensitivity β increases, even in the case, when proper donor centers doesn't participate directly in chemisorptional interaction. The increase of adsorption sensitivity is being realized due to the influence of the ionized donors through the diminishing φ_s on the free electrons' collective interaction of the semiconductor's subsurface region with atomic-molecular subsystem of its surface (see, please point. 1.2 in [1]).

If proper donors on the surface play the role of adsorption centers for external particles, so the increase of the donors' concentration could be the significant factor leading to the increase of the adsorption sensitivity of the *n*-type semiconductor to the adsorption of the acceptor particles.

In this case, besides the collective effects, the local interactions between the adsorption center and adsorbed particle should be taken into account despite the fact that they not in all cases could change the collective macro-parameters ρ and φ_s [1, 36–38].

The increase of the atomic-disperse (free) cadmium concentration along with the concentration and size of *Cd* clusters created on the surface of *CdSe* films cause the increase of their adsorption sensitivity to the oxygen [22–25]. In the absence of the cadmium accumulation on the surface e.g. at the starting point of the growth, the *CdSe* or *CdS* film, when the value of their ρ is high enough (Figure 3.3), the chemisorbed oxygen atoms are detected with the binding energy of (*O1s*) electrons is equal to E_B = 531.2 *eV* [22–27] (Figure 3.5). At the same time, the binding energy $Cd3d_{5/2}$ of the electrons has the value, characteristic for the *Cd* atoms in the structure *CdSe* E_B=405.3 *eV* (Figure 3.6). In this case, the *Se* vacancies are the main donors, which are compensated by the chemisorbed oxygen atoms and the *Se* vacancies could be considered as the centers of their adsorption [20, 22,23].

After the saturation of the surface with *Cd*, the free *Cd* atoms $Cd3d_{5/2}$ E_B = 531,8 *eV* (Figure 3.8) [1, 22–27, 34] are detected on the surface of *CdSe* and *CdS* films with the use of the *X*-ray photoelectron spectroscopy.

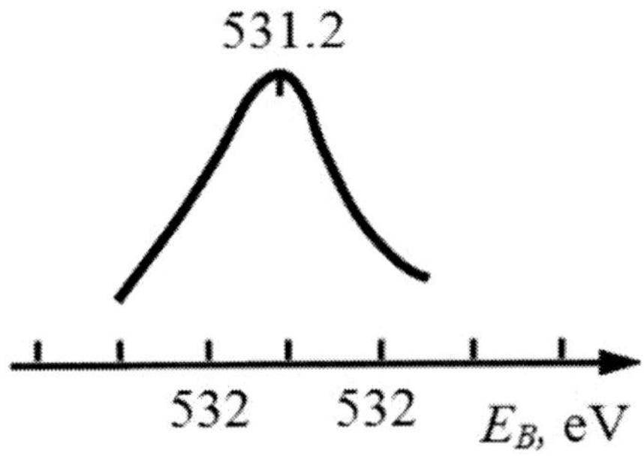

Figure 3.5. Spectrum of X-ray photo-electrons *O1s* of the starting (non-doped with *Cd*) of *CdSe* films surface.

The results of the papers [20–37] show, that the role of chemisorption centers on the surface of *n*-type semiconductor films could be played by the proper defects forming the donor energy centers in the forbidden zone on the surface and, besides, the chemisorption's form depends on the nature of the chemisorption on the surface of *CdSe* and *CdS* films. On the surface enriched with selenium vacancies (E_d =0.14 *eV*), the oxygen atoms *O1s* E_B = 531.2 *eV* are adsorbed mainly (Figure 3.5).

On the other side, the surface enrichment by the main donors – the excessive cadmium atoms (E_d = 0.3 *eV*) provides the chemisorption of the oxygen, mainly in molecular form (*O1s* E_B = 531.8 *eV*).

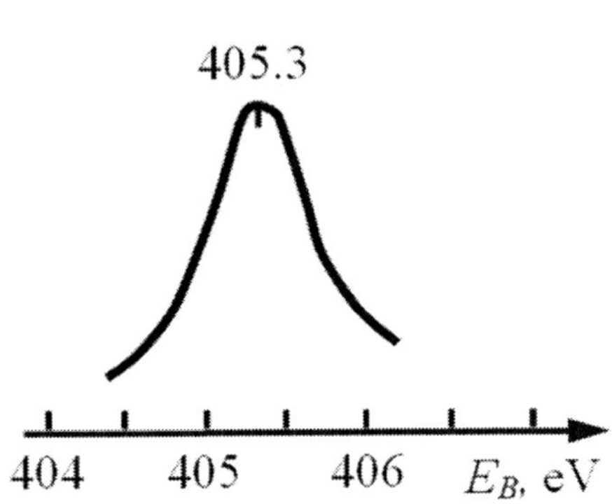

Figure 3.6. Spectrum of *X*-ray photoelectrons $Cd3d_{5/2}$ on the surface of *CdSe* film.

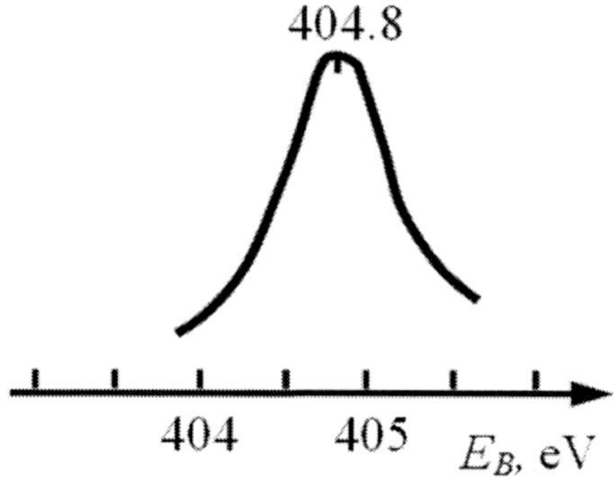

Figure 3.7. Spectrum of *X*-ray photo-electrons of cadmium $Cd3d_{5/2}$ on the surface of *CdSe* film, doped with cadmium.

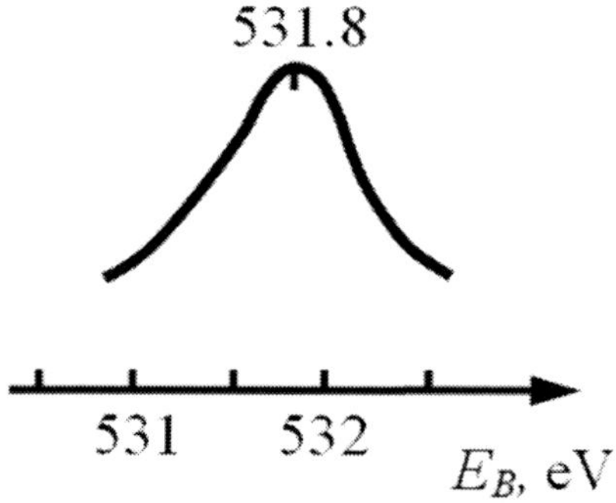

Figure 3.8. Spectrum of *X*-ray photo-electrons of oxygen *O1s* on the cadmium doped surface of *CdSe* film.

The observed mutually caused shifts in the spectra of *X*-ray photo-electrons of "free" cadmium on the surface of *CdSe* and *CdS* films along with the chemisorbed oxygen as well as the transfer from atomic to the molecular form of oxygen chemisorptions under the change of the main donors' nature from the selenium (or sulphur) vacancies to the inter-nodular cadmium atoms as well as the chemisorbed oxygen molecules' concentration increase accompanied by the increase of the cadmium atoms' concentration on the surface, have shown that the main centers of oxygen molecules' chemisorption are the cadmium atoms [20–37].

The data presented above show that the metallic atoms on the surface of *CdSe* and, *CdS* films, being the main donors, play, simultaneously, the role of main centers of oxygen molecules' chemisorption and the adsorption sensitivity' of these samples increase caused by the cadmium surface concentration increase. Concerning the film growth technology, one could state that the cadmium atoms' concentration may change in the growth process (Figure 3.6). But, it is impossible to reach the level of reproduction and monitoring of the adsorption sensitivity, using only existing simple methods of sensors' production (Figure 3.2). Thus, it's evident that it is necessary to realize the technology of sensors' production with the needed level of surface parameters' reproducibility using the surface doping of the semiconductor films.

References

[1] Smyntyna V. Electron and Molecular Phenomena on the Surface of Semiconductors. New York: NOVA Publishers; 2013.

[2] Smyntyna VA; Vashparov, YuA; Babinchuk, VS. Donor-acceptor defects and surface morphology of the films of cadmium selenide obtained in quasi-closed volume. Bulletin of the Academy of Sciences of the USSR. *Inorganic Materials*. 1985 vol. 21, 1293–1296.

[3] Турецкий, АЕ; Смынтына, ВА; Чемересюк, ГГ. Особенности фотоэлектрических свойств тонких пленок сульфида кадмия, полученных методом ЭГДР. В: 2-я Респ. конф. по фотоэлектрическим явлениям в полупроводниках : тез. докл..К: *Наук. думка*; 1982, 224.

[4] Golovanov, VV; Chemeresyuk, GG; Golovanov, VV; Ivanov, VF; Smyntyna, VA; Turetskiy, AE; Shmilevich, AM. Features of current transfer in thin films of cadmium sulfide due to electron phenomena on the surface. Physics, *Chemistry and Mechanics of Surfaces*, 1985 no. 4, 68–71/

[5] Smyntyna, VA; Moin, MD; Gerasutenko, VA; et. al. Influence of laser treatment on the adsorption interaction of cadmium sulfide films with oxygen. *Soviet Physics Journal*, 1990 vol.33 issue 3, 272-275.

[6] Shimada, M; Watanabe, H; Ishida, T. Manufacturing of fineline films by printing technique In: Proc. IMC. Tokyo: 1990 581–586.

[7] Nakao, K; [and others]. Giant micro-electronics technology using gravure printing techniques. In:Extended Abstract of the 1991 International *Conference on Solid State Devices and Materials.Yokohama:* 1991, 599–601.

[8] Stein, SJ; [and others]. Interconnection and packaging of advanced electronic circuitry. In: Proc. ISHM Conference. Orlando, Florida: 1991.

[9] Leppavuori, SA; [and others].A novel thick film technique, gravure offset printing, for the realization of fine line sensor structures In: *The Proc. Eurosensors Conf.* Budapest, Hungary: 1993,26–29.

[10] Исследование электронно-молекулярных явлений на поверхности полупроводниковых слоев и разработка основ физико-химических процессов в полупроводниковых сенсорах. Научный руководитель д-р физ.-мат. наук Смынтына, ВА. Заключительный отчет по теме № 161. Одесса:2002. № гос. регистрации 0100V002859. Инв. № 020311006656.

[11] Kazmerski, LL; Berry, WB; Allen, CW. Role of defects in determining the electrical properties of CdS thin films. *J. Appl. Phys.*,1972 vol. 43, 3521–3527.

[12] Bagociunaite, R; Pozela, J; Skucas, J; Skuceniee, A; Simulyte, A. CdTe pleveliu, legiruotu Cd elektrino laidumo tyrimai. *Lituanian Journal of Physics,* 1972 vol. 12, no. 6, 968-981.

[13] Karpovich, IA; Zvonkov, BN. The Hall mobility of electrons in the layers of sulfide and cadmium selenide. *Soviet Physics*, Solid State,1964 vol.6, no.11, 3392–3396.

[14] Волькенштейн, ФФ. Физико-химия поверхности полупроводников. М: Наука; 1973, 400 с.

[15] Киселев, ВФ. Поверхностные явления в полупроводниках и диэлектриках. М: Наука; 1970.

[16] Kolganov, VD; Mileshkina, NV. Adsorption and formation of a thin layer of gold coating on an atomically clean surface of GaAs. *Soviet Physics, Solid State*, 1985 vol. 27, no.9, 2580–2587.

[17] Крылов, ОВ. Катализ неметаллами. Л. : *Химия*; 1967, 240 с.

[18] Киселев, ВФ;. Зарифьянц, ЮА.Электроника на поверхности полупроводников и катализ В: Поверхностные соединения в гетерогенном катализе. *Проблемы кинетики и катализа. М: Наука;* 1975 т. 16, 221–237.

[19] Smyntyna, VA. Properties of the real surface of cadmium selenide thin films .Bulletin of the Academy of Sciences of the USSR. *Inorganic Materials,* 1982 vol. 18, no. 9 1479–1482.

[20] Smyntyna, VA. The chemisorption forms and the center nature of oxygen chemisorption on the CdSe thin film surface. II Nuovo Cimento, 1981 vol. 63B, no.2, 642–650.

[21] Golovanov, V; . Smyntyna,V; Mattogno, G. Oxygen Interaction of CdS - Based Gas Sensors with Different Stoichiometric Composition. *Sensors and Actuators*, B,1995.vol. 26–28, 108–112.

[22] Smyntyna, VA; Senkevich, AI; Gerasyutenko, VA; Latypov, SI; Dmitriev, AG. Influence of defects on the electron-molecular processes in the crystallites of cadmium selenide. *Russian Journal of Physical Chemistry* A, 1989 vol. 63, 1517–1521.

[23] Смынтына, ВА. Электрофизические, фотоэлектрические и электронно-молекулярные явления, стимулированные хемосорбционно-диффузионными процессами на границах полупроводниковых поликристаллических пленок селенида и сульфида кадмия : Дис. д-ра физ.-мат. наук. Одесса: ОГУ; 1988, 492с.

[24] Smyntyna, V; Golovanov, V; Mattogno, G. [and others]. Analysis of oxygen chemisorprition on CdS thin film surface. *Vuoto*, 1994 vol. 32, no.1, 86–88.

[25] Smyntyna, V; Kashiulis, S; [and others]. Chemisorbed oxygen forms on CdS thin surface. *Vuoto,* 1994 vol. 23, no. 2. 35–37.

[26] Kashiulis, S; Mattogno, G; [and others]. Adsorbed Oxygen Form on CdS Thin Film Surface. *Vuoto,* 1994 vol..23, no.1, 32–34.

[27] Mattogno, G; Viticoli, S; Kashiulis, S; Smyntyna, V. Analysis of Oxygen Chemisorprition on CdS Thin Film Surface .*Vuoto*,1994.vol. 23, no.1, 86–88.

[28] Александров, ЛН, редактор. Рост полупроводниковых кристаллов и пленок. Молекулярная лазерная эпитаксия. Распределение примесей и дефектов. Новосибирск : *Наука*; 1984, Ч. 1,160 с.

[29] Александров ЛН. Переходные области эпитаксиальных полупроводниковых пленок. М.: *Наука*; 1978, 272 с.

[30] Aleksandrov, LN. Transition regions in epitaxially grow semiconductor films and devices .*Thin Solid Films,* 1978 vol. 50, no. 1, 13–24.

[31] Smyntyna, V; Gerasutenko, V; Kashiulis, S; [and others]. The causes of thickness dependence of CdSe and CdS sensor sensitivity to oxygen. *Sensors & Actuators B.* 1994 vol. 19, no.1–3, 464–465.

[32] Смынтына, ВА. Изменение структурных, электрофизических и фотоэлектрических свойств в процессе роста и обработки пленок сульфида и селенида кадмия. *Фотоэлектроника.* 1988 вып. 2, 58–63.

[33] Герасютенко, ВА. Формирование свойств и состава пленок CdSe и CdS в процессе их роста и легирования: Дис. канд. физ.-мат. наук. Научный руководитель Смынтына, ВА.Одесса: ОГУ;1990, 201с.

[34] Smyntyna, VA; Gerasyutenko, VA; Korneeva SA. CdSe film surface property changes during growth. Bulletin of the Academy of Sciences of the USSR. *Inorganic Materials,* 1987 vol. 23, 1599-1601..

[35] Smyntyna, VA; Vashpanov, YuA; Babinchuk, VS. The changing nature of the major donors and the surface morphology of CdSe layers with increasing thickness and the condensation temperature // Bulletin of the Academy of Sciences of the USSR. *Inorganic Materials*, 1985 vol. 21, 1134-1137.

[36] Вашпанов, ЮА; Смынтына, ВА. *Адсорбционная чувствительность полупроводников.* Одесса : Астропринт;2005,216 с.

[37] Голованов, ВВ. *Характеризация поверхностно-активных веществ* $Sn_xW_yO_z$, *CdS* $(Se)_{1-x}$, In_2O_3 *Дис. д-ра физ.-мат. наук.* Научный консультант Смынтына., ВА.Одесса: ОГУ;1998, 500 с.

[38] Morrison S. Roy. *The chemical physics of surfaces*. New York: Plenum Press; 1977, 415 p.

Chapter 4

The Influence of Sensor Surface Alloying on Adsorption Sensitivity

Abstract

The fourth chapter describes the effects and phenomena caused by doping the surface of materials for gas sensors. The changes of the adsorption sensitivity of semiconductors surface by doping and variation forms of chemisorption and the binding energy of the adsorbed particles to the center of adsorption are considered. Unusual dual effect (donor and acceptor) of chemisorption of the same type of particles on the same semiconductor is discussed. The reasons of changes in the kinetics of chemisorption under the influence of doping the surface are analyzed.

4.1. Effects, Caused by Alloying of Gas Sensor Surface

Surface alloying is one of methods that are used for varying of semiconductor surface properties. It gives significant impact on electronic properties of semiconductors [1, 2] and on their ability to interact with gas environment [3]. Surface electronic potential [4, 5], trapped charge concentration on surface states [6], mobility of electrons and holes within surface region [7, 8], electron work function [9], lifetime of nonequilibrium charge carriers [10] and other parameters are changing because of alloying applying. In case of alloying by metals the changes of surface electronic properties of semiconductor depend on the ratio of electrochemical potentials of used metal and semiconductor material [11]. Metal impurities, which have more positive electrochemical potential than semiconductor, are deposited on the surface in elementary form. Impurities with more negative electrochemical potential are forming compounds on the surface, whose influence on the surface electronic properties is poorly studied.

Doping of the surface can be done in many ways, including precipitation of alloying elements in the liquid phase [11], materials thermal evaporation in vacuum [12]. In first case, the semiconductor surface for a few minutes is brought in contact with the solution containing dopants with concentration in range $(10^{-5} \pm 10^{-1})$ *g-at/l*. In this case, the concentration of the applied atoms cannot be less $10^{17} m^{-2}$ [11]. In second case, the material is absorbed from the vapour phase, and its concentration can be changed in a wide range of values depending on process conditions. In [13] adsorption of oxygen radicals at a concentration 10^{-5} % of a monolayer value was registered by the method of measuring the electrical conductivity, so that means, concentration of radicals on the surface was $10^{12} m^{-2}$.

It is also known that the impurity can be applied to the surface of the semiconductor by spraying solution from the nozzle [14, 15], by ion implantation [16], by electrodeposition from the liquid phase [17], by impregnation of materials by subsequent solutions [18], by the optical breakdown of gas using laser radiation [19], by high-power laser irradiation of surface (laser ablation) [20] and other derivative techniques.

Influence of surface alloying (SA) of metal salt solutions on the electronic properties of semiconductor surfaces studied in [21] where it was shown that at the semiconductor surface the microparticles of different size are formed [22]. The dispersion degree of particles and their structure depend not only on the properties of the adsorbed material, but also on the physical and chemical properties of the semiconductor surface [23].

Oxide layer on the silicon surface prevents adsorption of *Au* and *Ag* in the elemental form. Surface alloying from the liquid phase, that dissolves the oxide film, improves condensation conditions and leads to the creation of a dense structure [24]. In this case a significant increase in the degree of coagulation of impurities islands and growing tendency to the formation of elongated structures are observed [24]. Coagulation prevents the obtaining of high-dispersive impurity structures, in which case there is the most effective influence of alloying elements on the surface properties is detected [23]. Coagulation can take place during slow or sudden movement of islands that contribute by the electrostatic forces of attraction between the particles.

To obtain and save a high degree of dispersion of the particles on the surface it is needed to reduce the mobility of the alloy particles, for example, by increasing their connection with the semiconductor surface [23]. For this purpose, the semiconductor is applied with intermediate substance that forms a fairly strong bonding with both semiconductor surface and adsorbed dopant. For example, transition metal ions can be used as an intermediate material, which, on the one hand, form an ionic bond with ion crystallites, and on the other - form a covalent bond with alloy particles due to the remaining d-electrons.

It was found in studies of ammonia oxidation on *Pt* clusters that the binding energy of oxygen with clusters of small size is greater than with larger clusters [25]. Therefore, the oxygen activity in ammonia oxidation reactions with using a platinum catalyst in the form of small clusters is significantly lower than in the case of using the large clusters. This makes it possible to control the speed of technological process.

Doping of silicon surface by atoms of silver, indium, platinum, gold, zinc, manganese leads to a significant change in its electronic properties. Depending on the method of doping and the individual properties of the adsorbate both as an increase and a decrease in the density of surface states and surface recombination rate can be observed [26].

Semiconductor surface can be considered as a disordered system, which quasicontinuous spectrum of fast surface states, for example in the case of silicon, is caused by the violation of the translational symmetry of the atomic potential on the semiconductor–oxide border. The causes of these disorders are significant variations of bond angles between adjacent SiO_4 tetrahedrons and edge lengths of these tetrahedrons [26]. Because of the considerable degree of ionicity of *Si-O* bonding, tetrahedron deformation changes the effective charge of the atoms of a tetrahedron, which in turn creates fluctuating surface potential. During surface alloying by metals Zn^{2+}, In^{2+}, Mn^{2+}, In^{3+} the complexes $Me^{n+}O^-$ are formed on the surface.

Having a vacant *p*- and *d*-orbitals, these metals compete with silicon for the electron density of the oxygen atom, entering with the latest in additional donor-acceptor interaction. Therefore, even if the doping does not change the structure of the surface oxide layer, the changes in the chain of *Si-O* adsorption of metal atoms will lead to a change in the charge of the atoms, which, according to the model of disordered structures, accompanied by a change in the spectrum of surface states.

The change in the surface barrier by surface alloying leads to the changes in position of the Fermi level in the semiconductor band gap. This opens up the possibility of controlling the surface properties both in terms of increasing its activity to the gas environment, and from the point of view of passivation [27].

In the study of the influence of alloying metals from the vapor phase in a vacuum on the properties of semiconductor surfaces, it was found that, regardless of the donor-acceptor properties of the impurity atom in the bulk of the crystallite and its chemical nature in the case of *n*-type semiconductor, there is a reduction of its work function X_s [28] and the surface potential φ_s [29]. All metals that were used as dopants, show themselves on the surface as a monovalent electron donors [30]. Magnitude of the changes in the work function and in the surface potential value is different for different materials. In the line of investigated metals (indium, copper, silver, gold) the most active influence on the surface properties of germanium, silicon, gallium arsenide, cadmium selenide films have indium atoms [8, 12, 14, 22], which alter the surface band bending in these semiconductors for a value of several kT units respectively [31]. Gold is the least active. The greatest changes in the surface potential, work function and electrical conductivity of the semiconductor are happening in the initial stages of metals adsorption at concentrations $N_s > 10^{18}$ m^{-2}. A further increase of particles concentration on the surface is accompanied by a significant decrease in the rate of change φ_s, X_s and electrical conductivity Ω. When $N_s > 10^{18} m^{-2}$, there is a further reduction in the work function, which is not related to the change of the surface barrier, and caused by the formation of a dipole layer, which consists of adsorbed particles and adsorption centers [31].

The conductivity increase due to doping of the surface by metal atoms is connected with the increase in the free electrons concentration in the conduction band for *n*-type semiconductors as a result of electron transition to the conduction band from surface level of the adsorbed metal atoms, which serves as a donor [30]. In this case, the metal atom becomes ionized, and it remains positively charged on the surface. The adsorption of metal atoms with forming of positive ions on the surface of the n-type semiconductor causes significant changes φ_s, X_s, Ω. In case of increasing N_s the concentration of positive ions on the surface increases, and this makes it difficult the ionization of the incoming metal atoms and it leads to a decrease in the rate of surface parameters changes [30]. Therefore, in the later stages of adsorption centers filling further ionization of adsorbed atoms may stop because of adsorption sites number limitation, which can be involved in chemisorption of the metal atoms.

The lowering of the potential barrier during the surface alloying by metals causes an exponential decrease in the time constant of the electronic exchange between the surface and the volume and greatly increases the intensity of the electron capture by the surface states [32]. In the case of electron-depleted surface of cadmium selenide fast surface states (SS) are almost absent near the Fermi level (time constant $< 10^{-3}$ s), but then after zones rectification (as a result surface alloying) their density comes to 10^{17} m^{-2} eV^{-1}. These SS are playing role of effective electrons capture centers.

A gradual recovery of semiconductor surface original (before doping) parameters is observed after SA by metals to a concentration $N_s \approx (1 \div 5)10^{18}$ m^{-2} [33]. The surface relaxation times of getting to its original condition vary from seconds to hours, depending on the concentration of impurity atoms and the nature of the semiconductor.

Relaxation phenomena are associated with the establishing of adsorption-desorption equilibrium in semiconductor – boundary layer – atmosphere system that was disturbed by surface alloying [33]. The effective electronegativity of the layer decreases because of the interaction between metal atoms and boundary layer, and this should lead to the enrichment of the surface region by electrons, if the electronic exchange with the semiconductor takes place. Reducing the concentration of free electronegative components of the boundary layer leads to disruption of the electron-ion and adsorption-desorption equilibrium in the system. Restoring the balance due to additional adsorption from the electronegative gases surrounding atmosphere is the cause of the surface state relaxation after doping.

Thus, when the surface alloying is applied, both the semiconductors surface electronic properties and its interaction with the surrounding gas are changing. The influence of SA on the surface properties of metals and semiconductors is used mainly to determine the mechanism of the formation of metal - semiconductor electric contact [24].

At the same time, semiconductor surfaces doping, not only by metals but also by other chemical elements has considerable research interest, because imposed atoms are changing not only the electronic properties of the metal, but also electrical activity of the surface during contact with the gaseous medium, and, therefore, this is giving possibility to determine the role of the surface atoms in the physical and chemical processes occurring on the semiconductors surface.

Changes in electrochemical and adsorption properties of the surface at SA are occurred as the result of Fermi level position changing in the band gap on the surface (collective mechanism), and because of the direct participation of the impurity atoms in the adsorption process as adsorption centers or its blocking elements (local interaction) [34-37] . In this case, the collective and local mechanisms of impurity influence may occur simultaneously, and they can act either in the same direction, or they can create the opposite effects. This greatly complicates the analysis of SA results and interpretation of phenomena, caused by SA.

4.2. Alloying Influence on Surface Adsorption Sensitivity

4.2.1. *CdSe* Films

It is assumed in the discussion of the alloying processes that doping addition is forcibly entered in the sample volume or on the surface by special methods and technologies. However, the process of semiconductor material doping can take place in a work cycle arbitrary, as a result of processes that are not provided by technology. For example, during thin film sensor manufacturing in vacuum, components of the residual atmosphere interact with the particles of the evaporated semiconductor material and can be incorporated into the structure of produced films by creating replacement or implementation defects, or can mostly dissipate some component of evaporated material on its way to the substrate, and so thereby to create vacancies in the lattice of the thin film structure.

Oxygen is the most active gas in the residual atmosphere under vacuum bell. Increase of its pressure in the technological volume from 10^{-2} *Pa* to 1 *Pa* leads to changes in the cadmium selenide films structure from cubic (Figure 4.1, a) to the hexagonal (Figure 4.1, b) [38]. There was not detected formation of oxide phase in *CdSe* film structure in the equipment sensitivity limits [39]. Specific resistivity decrease that is observed in this case (Figure 4.2, curve 1) is caused by the donor defects accumulation in the structure of the film during the growth, such as selenium vacancies or excess cadmium atoms. These defects, acting as chemisorption centers, lead to an increase in the sensitivity of the adsorption surface (Figure 4.2, curve 2) [38, 39].

Several processes are taking place simultaneously, as the oxygen pressure in technological vacuum volume increases. Selenium is evaporating in the form of Se_2, Se_4 molecules that are considerably scattered by residual atmosphere components.

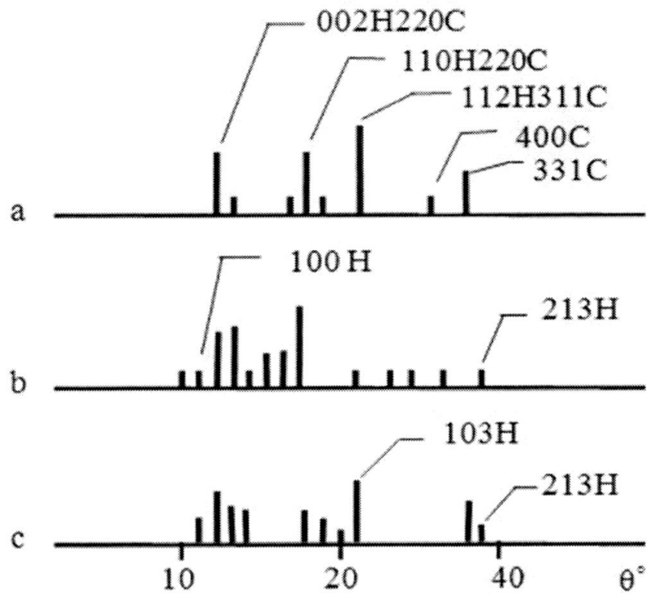

Figure 4.1. Bar diagrams of diffractograms of *CdSe* films deposited at different oxygen pressures in the process chamber: $1{,}1\times10^{-2}$ *Pa* (a), 4×10^{-2} *Pa* (b), $1{,}2\times10^{-1}$ *Pa* (a).

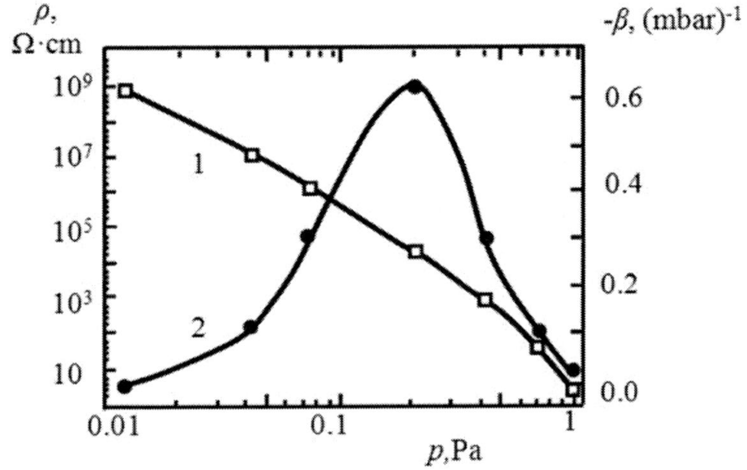

Figure 4.2. Dependence of specific resistivity of cadmium selenide layers (1), and their sensitivity to oxygen adsorption (2) on the pressure of oxygen in the process chamber.

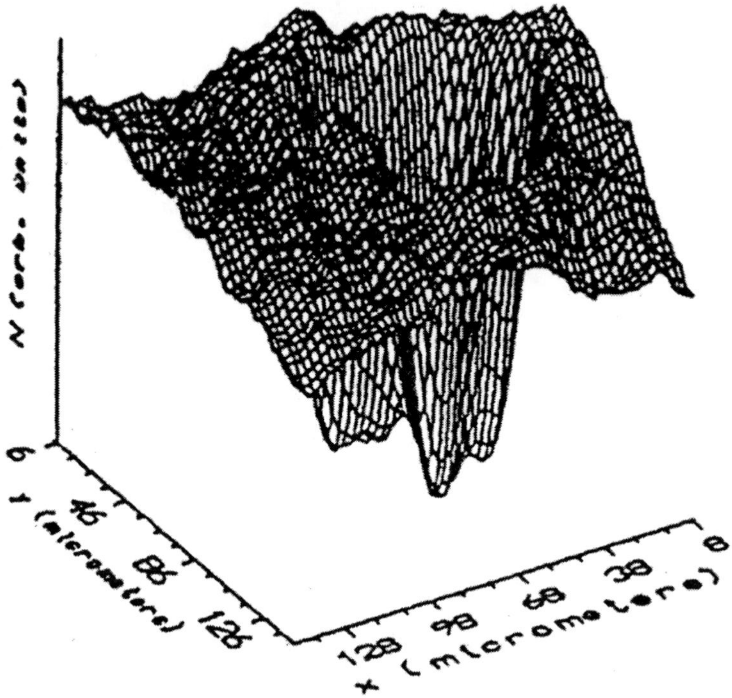

Figure 4.3. Image of the film Pt/SnO_2 surface with "holes" in the porous layer of platinum with thickness of (5-10) Å [42].

Due to this, an excess of cadmium in crystal structure appears by forming selenium vacancies.

As a result, first, due to the acceleration of the crystal lattice strain its cubic structure, that has little tolerance for defects, is transformed into the hexagonal modification. Second, the accumulation of donor defects in the film structure in the form of excess atoms of cadmium and selenium vacancies with increasing O_2 pressure in the technological vacuum volume leads to adsorption sensitivity increment [1, 38].

The accumulation of a large concentration of excess cadmium atoms in the structure of the film can lead to the ousting them during the process of recrystallization on the surface, where they form a teardrop shape macrodefects, between which coalescence can occur. As a result, the concentration of chemisorption centers decreases and adsorption sensitivity of films obtained at high pressures of residual oxygen is reduced (Figure 4.2, curve 2) [1, 40, 41].

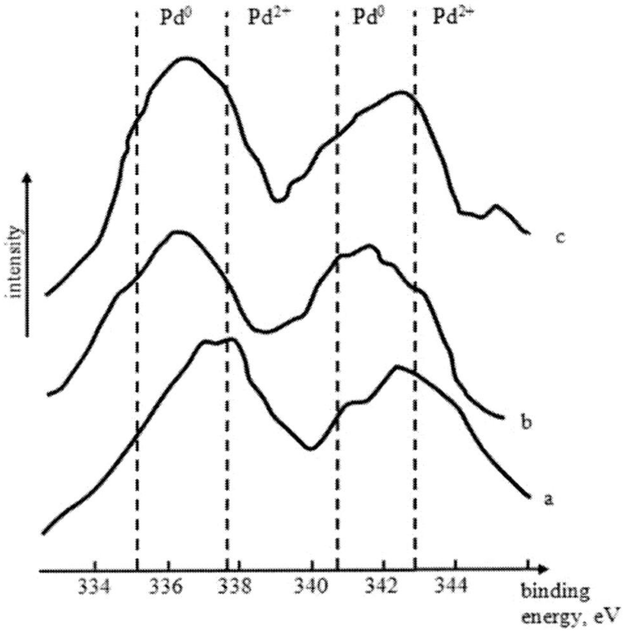

Figure 4.4. *X*-ray photoelectron spectra of *Pd* 3d lines in the $Pd/Pt/Al_2O_3$ sensor, annealed a) in air at 770 K, b) in the methane-air mixture (1% vol. CH_4) at 600 *K* and c) at 720 *K* during 1 hour.

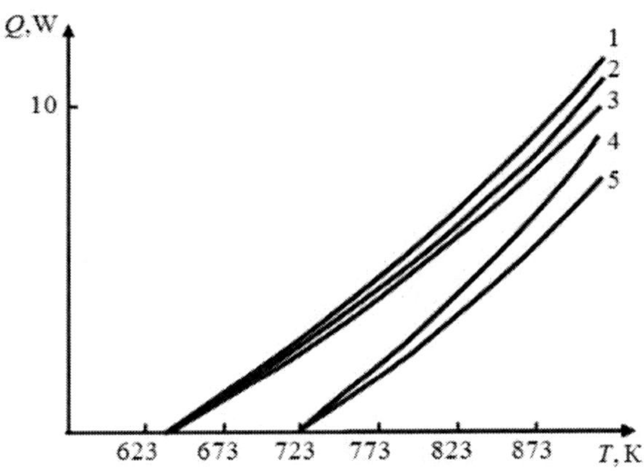

Figure 4.5. Curves of the catalytic activity *Q* of In_2O_3 sensors (1, 2) and Pt/In_2O_3 (3-5), measured in methane-air mixture (concentration of CH_4 1% vol.) At temperatures 1, 5 – 1270 K, 2, 4 – 1070 K; 3 – 770K.

4.2.2. Metal Oxide Films

Admixture which improves gas sensors adsorption sensitivity for various purposes is introduced as a supplement to the initial mixture during preparation of sensing element processing [42-48]. Insertion of 5% $PdCl_2$ in the mix for the SnO_2 gas sensor preparation greatly increases its sensitivity for methane [44]. To increase the sensitivity of SnO_2 gas sensors for CO it is effective to cover the surface by layer of ultra-thin Pt (d = 5-10 Å), with following annealing it to the surface [41]. Pt layer in the initial state (before annealing) is very heterogeneous and porous, as can be seen from the measurements of the Auger spectra, it contains irregular shaped "holes" with (30 ÷ 80) micrometers diameter (Figure 4.3). Determinative importance in sensors sensitivity for CO increasing by Pt using is evidenced by the fact that the Pt/SnO_2 and Pt/SnO_2: Sb samples have much greater adsorption sensitivity than SnO_2 or SnO_2: Sb sensors.

Insertion of Pt and Pd atoms to Al_2O_3 gas sensors significantly enhances their sensitivity to catalytic interaction with methane [43]. The main reason for enhancing the catalytic activity of $Pd/Pt/Al_2O_3$ structures during CH_4 catalytic oxidation is the reduction of Pd^+ and Pd^{2+} to Pd^0 form by the presence of Pt (Figure 4.4).

Pt dopant decreased the starting temperature of metal catalytic oxidation on Pt/In_2O_3 gas sensors from 720 K to 640 K (Figure 4.5). In this case, at temperatures below 640 K Pt/In_2O_3 sensor signal is due to CH_4 chemisorption on the semiconductor surface, and at temperatures above 643 K – due to catalytic oxidation of methane.

4.2.3. *CdSe*: *In* Films

Film surface alloying of *n*-type A_2B_6 compounds, mainly *CdSe* and *CdS*, by metal atoms leads to an increase in their adsorption sensitivity regardless of the nature of metals (*In, Cu, Ag, Au, Cd, Ga*), and to acceptor chemisorption of gas molecules, especially oxygen, the most effective dopant in this case is indium [1, 40-42, 49].

There is a reduction in the work function and sensor surface potential, regardless of the *n*-type semiconductor nature and properties of impurity materials used in doping of the semiconductor surface by spraying in a vacuum [50-52], and all the studied impurities behave as monovalent donors [53].

Within the resolution of the used instruments, it was found that at the early stages of doping indium atoms are not in isolated from each other, but form isolated clusters (Figure 4.6, a). Then "chains" of these clusters are formed, this leads to appearance of grid structure on the surface of cadmium selenide films (Figure 4.6, b), "cells" indium filling subsequently leads to the formation of a continuous *In* film (Figure 4.6, c) [1, 40, 54].

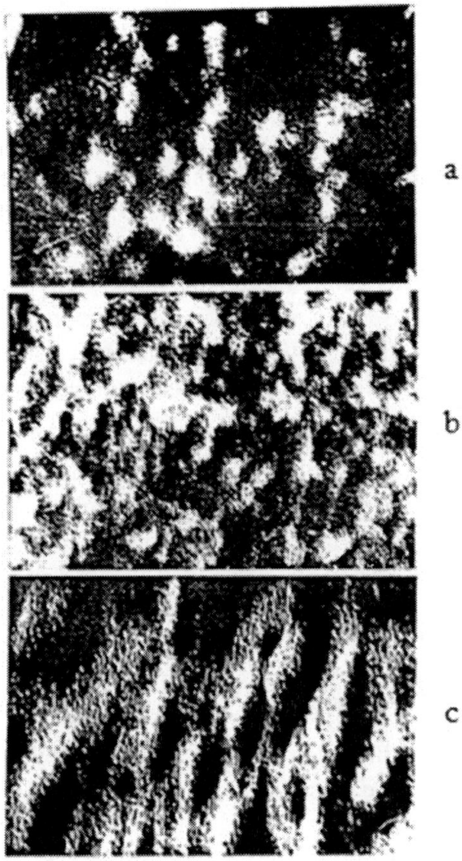

Figure 4.6. The change of surface morphology of cadmium selenide films, depending on the time of doping by indium. $T_{In} = 990$ K, $T_{substrate} = 300$ K. The doping time: a) 10 *s* (× 4000), b) 20 *s* (× 4000), c) 60 *s* (× 7800).

Alloying *In* atoms create donor levels that are situated above the Fermi level, and are on the surface in the form of ions [55], their positive charge partially compensates the negative surface charge Q^{-}_{SS}. In this case, as the concentration increases, the decay of the height of the surface potential barrier for electrons is observed, which leads to a significant increase of β because of its heavy dependence on Y_S [56], the relative decrease surface potential layer of cadmium selenide of the measured by Kelvin method on (10÷15) kT units was observed at room temperature when the concentration was up to 10^{18} m^{-2} [1, 41].

The appearance of noticeable sensitivity to oxygen, measured at 333 K, 343 K, 353 K, 363 K was experimentally observed (Figure 4.7, curves 1-4) for $CdSe{:}In$ films at concentrations N^s_{In}= $10^{17}m^{-2}$, which increases with the growth and is reduced after reaching a maximum (N^s_{In} = $5\times10^{18}m^{-2}$) [52].

Increase of N^s_{In} above $5\times10^{18}m^{-2}$ is accompanied by an increase in the concentration of electrons in the conduction band up to degeneration. In this case, the concentration of electrons trapped by oxygen is less than the concentration of free electrons remaining in the conduction band, and the value of β decreases.

As the measurements temperature increases from 323 K to 373 K, the ionization degree of the main bulk of the semiconductor donor states is growing up to 0.14 eV and 0.3 eV. This dramatically increases the number of electrons in the conduction band and, as a consequence, the value of the bulk conductivity σ_{CB}. According to the results [56], the σ_{CB} growth should significantly reduce the level of sensitivity. In addition, as the temperature increases, the probability of desorption also increases exponentially, which should lead to decay of chemisorbed oxygen molecules concentration and, consequently, to decrease of β.

However, the experimentally observed increment of β with temperature increasing from 323 K to 353 K (Figure 4.8) is due to the increase in the concentration and depth of oxygen-origin electron capture centers on the surface of semiconductor films [1, 41].

The use of "pure" cadmium selenide as a sensitive element of gas analyzer is rather complicated by the presence of a potential barrier at the surface of the film, which makes it smaller the probability of electron capture from the conduction band to the level formed by chemisorbed oxygen at room temperature. Increment of the rate of electron capture by the oxygen level occurs with temperature increasing. However, in this case the rate of oxygen diffusion into the bulk of films increases [41], and this causes an undesirable raise of the inter-crystalline barriers [1]. Surface alloying of cadmium selenide by indium atoms lowers the surface potential barrier, because indium is a donor material for A_2B_6.

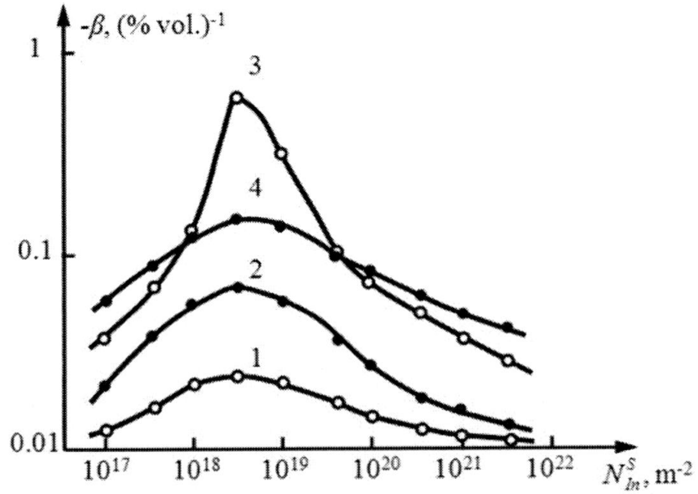

Figure 4.7. Dependence of adsorption sensitivity level on the level of surface doping by indium for cadmium selenide films. β is measured at T = 333 K (1), 343 K (2), 353 K (3), 363K (4), for c = 20% of the oxygen in the mixture of N_2, O_2.

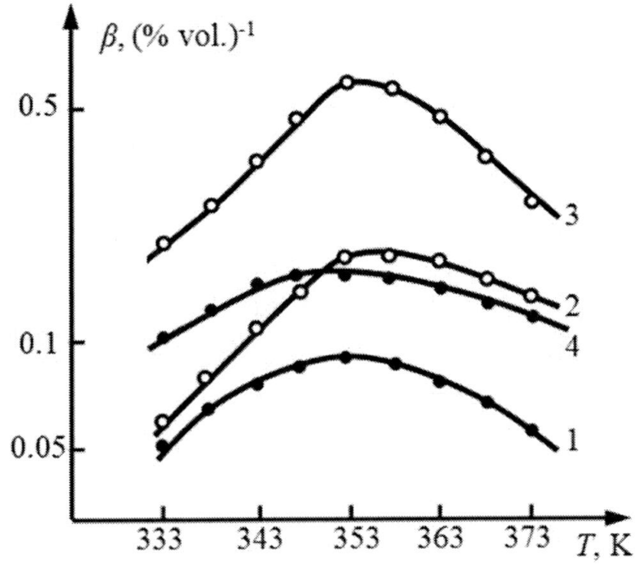

Figure 4.8. The temperature dependence of adsorption sensitivity β of *CdSe:In* films, doped by indium atoms using *In* concentration: 10^{13} cm^{-2} (1), 5×10^{13} cm^{-2} (2), 10^{14} cm^{-2} (3), 10^{15} cm^{-2} (4).

As a result of indium surface doping of *CdSe* films with 1 *μm* and 5 *μm* thickness in a single technological cycle of deposition in a vacuum the increase of adsorption sensitivity *β* up to (40-50)% filling of sensor surface by indium was found, regardless of cadmium selenide film thickness. Further enhancing of the degree of sensor surface filling by the alloy leads to the loss of its sensitivity, that is caused, ultimately, by shorting the sensor by indium film [1, 40, 41, 52].

Cadmium selenide films with thickness of 1 *μm* doped by *In* atoms at entire range of used dopant concentrations have a much greater adsorption sensitivity in comparison with *CdSe* sensors of 5 *μm* thickness. It is still not clearly identified – whether it is the size effect responsible or the changes in the defect level and the composition of surface defects of *CdSe* microcrystal planes, which are increasing during film thickness enlargement, have the major influence here.

Film surface doping by indium leads to increase of their sensitivity to oxygen adsorption, first, by the fact that the concentration of oxygen adsorption centers increases, which was detected by *XPS* method [1, 41]. Second, because the indium atoms, being shallow donors, are ionized even at room temperature [55] and their positive charge compensates the negative charge of the biographical surface, and that causes reduction of the surface potential barrier. This increases the probability of hitting an electron from the conduction band in the bulk to the surface levels of adsorption origin.

Regardless of the *CdS*e crystallites size, the most significant increase in the adsorption sensitivity is observed at that time of doping, when *In* microclusters are formed and their number is relatively rapidly increasing (Figure 4.7, curves 1, 3, 4). And simultaneously with the In clusters chain structures formation of the film surface, the maximum of adsorption sensitivity is recorded. Further formation of the "net" and its cells filling with indium leads to the reduction of the adsorption sensitivity (Figure 4.6). These data show that the increase in the adsorption sensitivity is due to the doping of the film surface of cadmium selenide by indium atoms, and regularities in the changes of its value are caused by features of the formation and evolution mechanism of indium clusters.

4.2.4. *CdSe: Se* Films

As a result of surface alloying by selenium by vacuum evaporation it on the surface of *CdSe* sensitive films in the concentration range between 10^{17} m^{-2}

$< N^{s}_{Se} < 5\times10^{18}\, m^{-2}$ with its growth is a significant reduction in conductivity Ω of *CdSe* layer, and also decrease of the sensitivity to oxygen chemisorption are observed. Sensitivity β, measured at $T = 353\ K$, was decreasing with 10% vol. of oxygen concentration increment (Figure 4.9). Layers of *CdSe:Se* did not respond to oxygen concentration changes in the test chamber in range from $5\times10^{18}\, m^{-2}$ to $10^{19}\, m^{-2}$.

At concentrations greater than monolayer, after films heat treatment in 10^{-5} *Pa* vacuum at a temperature $T = 453\ K$, layers demonstrated p-type conductivity at room temperature.

Temperature dependence of the dark current has shown that this dependence is characterized by activation energy of 0.64 *eV*. In *p*-type films of *CdSe: Se*, this activation energy is associated with cadmium vacancies in the structure of the semiconductor [55].

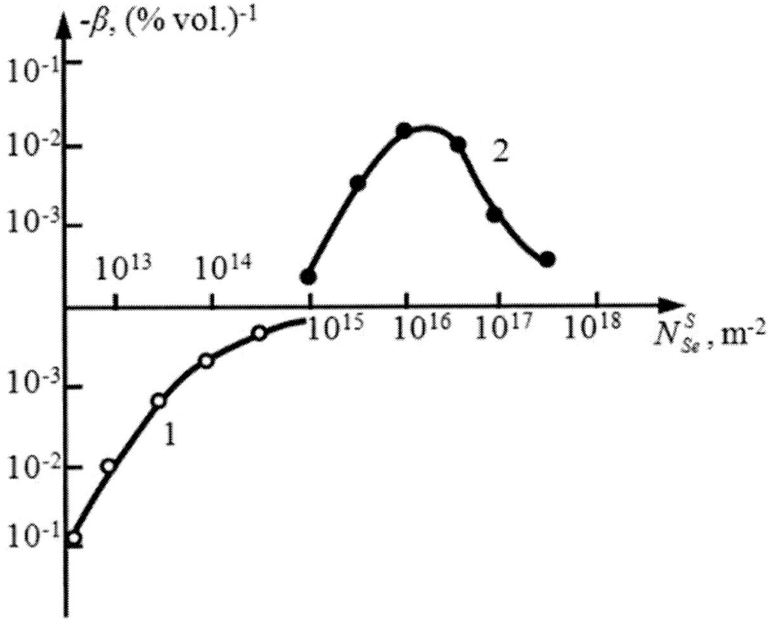

Figure 4.9. The dependence of the adsorption sensitivity β on the degree of cadmium selenide surface doping by selenium atoms. *CdSe:Se n*-type (1), *CdSe:Se p*-type (2).

Vacuumization of measuring chamber was accompanied by a decrease of conductivity of *p*-type *CdSe:Se* films, that were obtained after heat treatment. Oxygen insertion into the measuring chamber increases the room temperature conductivity of films. β of all p-type *CdSe:Se* films, measured at room temperature, has a positive sign, which is consistent with the calculation of β (Y_S) for this case in [56]. The calculation indicates that β becomes positive in case if *p*-type layer appears on the surface of *n*-type *CdSe* films.

Reduction in sensitivity β with increasing of N_{Se}^{S} for n-type *CdSe:Se* films can be associated with an increase in the band bending up due to surface donors charge compensating by acceptors (*Se*) and also due to appearance of the negative charge on the surface. It is shown in [1, 56] that cadmium selenide surface doping by selenium atoms decreases the adsorption capacity of the semiconductor surface, and increases the binding energy of oxygen molecules with the layer surface. Reducing the number of oxygen ions on the surface as a result of cadmium selenide doping by selenium should lead to a decrease of the charge, localized on the adsorbate, the overall surface charge and, thus, to reduce the sensitivity β. The fact of experimentally observed decrease in the sensitivity β, suggests that the increase in the band bending up at the surface and adsorption capacity reduce has a stronger effect on β, than the increase of the binding energy of oxygen to the film surface.

The resistance of a *p*-type cadmium selenide semiconductor film increases monotonically with time, and most significantly, if oxygen in the camera was present. Change in conductivity was accompanied by a decrease of the adsorption sensitivity with increasing of oxygen exposure time. After a stay of semiconductor films in oxygen atmosphere for 3-5 hours, layer resistance was greater than $10^9\Omega$, and it almost did not respond to variations of oxygen concentration.

CdSe:Se layer had a reddish tint immediately after the deposition. The layers turned gray after the interaction with oxygen, when the resistance of films increased to $10^9\ \Omega$ and films lost sensitivity to oxygen. It is known that the red color corresponds to the monoclinic selenium, and gray – to hexagonal [57], i.e., *CdSe:Se* films tint is determined by the crystal structure of microcrystals (clusters) of selenium on the surface of doped cadmium selenide films.

It follows that the observed increase in resistance and a time decrease in β occurs simultaneously with recrystallization of selenium structure on the semiconductor surface. Rate of increase in resistance of the cadmium selenide films is higher, when samples are put in an atmosphere of oxygen, indicating that oxygen promotes the recrystallization of selenium.

Change in electrical conductivity, which occurs as a result of the chemisorption-desorption interaction of oxygen with the surface of *p*-type cadmium selenide films, can be connected not only with the capture of a free carrier, but also with the redistribution of the generated charge of biographical origin, which takes place in the local interaction between own active centers on the semiconductor surface and chemisorbed oxygen. These active centers can be formed by negatively charged cadmium vacancies or selenium atoms on the layer surface. For example, changes in the electrical conductivity may be due to the changes in the charge state of selenium atoms from Se^{+4} to Se^{+6} in the group SeO_3, SeO_4, in which oxygen competes for the electron density with selenium atoms. The changes in the charge state of selenium atoms, in turn, can affect the rate of recrystallization of selenium on the film surface [52].

4.3. Influence of Sensor Surface Alloying on Chemisorption Type and on Binding Energy of Chemisorbed Particles

Adsorption properties of surface are varying because of its interaction with doping atoms during surface alloying. The simplest effect here is the increment of chemisorption centers concentration by increasing of dopant atoms surface concentration.

However, this is not the only phenomenon that accompanies the alloying process, even at the stage when the doping atoms can still be considered in atomic dispersed state on the surface.

If the alloying metal is a donor for the certain type of *n*-type semiconductor, the local energy level, generated by doping atoms, will be situated on the surface in the band gap above the Fermi level [1]. In this case the doping atoms donate their electrons to the conduction band, and thus the positive charge, formed on the surface by the ionized ligand atoms, compensates biographical negative surface charge and, thus, reduces the surface potential barrier. This increases the probability of electrons capture from the conduction band to capture levels of adsorption origin, that are formed by acceptor-type chemisorbed particles below the Fermi level at the surface. As a result, the sensitivity of the adsorption surface increases, first, because of the growth of concentration of free electrons, that are able to participate in chemisorption process within the conduction band (including the electrons supplying there from ligand atoms), and second - due to the increased probability of electron capture from the conduction band to the acceptor level of the chemisorption origin, by reducing the height of the surface potential barrier [1].

Chemisorbed particles are sensitive to the nature of the adsorption centers. Therefore, with successful selecting of the ligand nature, the concentration growth of its atoms on the surface leads to an increase of the chemisorption centers concentration for particles of a given gas type and to the adsorption sensitivity increase of the surface to its chemisorption [1].

If the task is to increase the surface adsorption sensitivity to the particles of donor gas, the surface must be alloyed by those metals, which atoms create acceptor levels in the band gap on the surface of *n*-type semiconductors. Localization of electrons at these impurity-origin electron capture levels increases the surface potential barrier. This in turn creates the conditions for the increment of material adsorption sensitivity to the chemisorption of the donor gases particles.

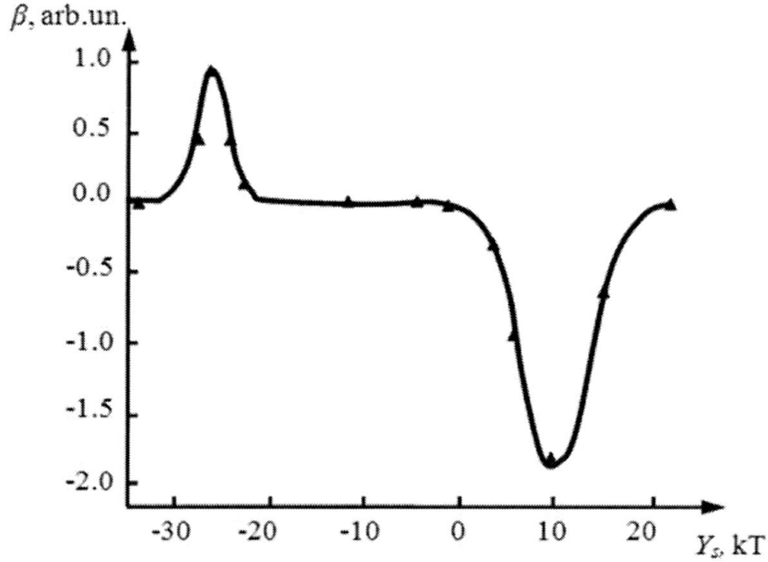

Figure 4.10. Dependence of β on the value of the surface potential Y_S ($E_g = 1.5$ eV).

In order to obtain semiconducting samples that could be used as an adsorption-sensitive elements (ASE), also called as the gas sensors, it is necessary that their electrical conductivity was determined by the surface properties, i.e., either the bulk conductivity should be less than the surface conductivity, or the thickness of the sample should be about Debye length.

The successful introduction of adsorption sensitivity definition, especially for the latter case, [1, 49, 56]

$$\beta = \frac{1}{\sigma} \cdot \frac{d\sigma}{dc} \tag{4.1}$$

or

$$\beta = \frac{1}{\sigma} \cdot \frac{d\sigma}{dp} \tag{4.2}$$

(σ - electrical conductivity of the semiconductor, c, p – concentration and pressure of sertain type of gas, respectively) allowed to establish theoretically that β depends mainly on the surface potential, and hence, on the parameters that determine its value. Theoretical analysis [1, 49, 56] showed that, due to this, the possible submission of β through the changes in the concentration of free electrons n and mobility μ, the expression for $\sigma = en\mu$ contains, has practical value only in certain, specific cases.

From numerical calculations [1, 49, 56] it follows that two extreme regions of adsorption sensitivity should be present at the dependence of β on surface potential Y_S. For example, as it is observed in the case of n-type semiconductor sensor surface enrichment by donors to acceptor particles chemisorption or in the case of the inversion of surface conductivity type for chemisorption donor particles (Figure 4.10).

Semiconductor surface chemisorption of the atoms and molecules of the same gas is to be regarded as the adsorption effect of two different particles, as they are characterized by different values of both chemical and physical characteristics. If we consider the chemisorption of oxygen atoms and molecules on the surface of cadmium selenide film, by virtue of differences in their electronegativity (O - 1,4 eV, O_2 - 0,9 eV [1]), it is expected that substantial differences in the results will be observed.

During chemisorption of O_2 molecules the electron capture centers in the band gap at the surface are formed, the energy levels are located at 0.75 eV below the conduction band and the concentration of trapped electrons can be up to 5×10^{16} m^2 [53]. The cross section for electron capture is (3-5) $\times10^{-21}m^2$.

Atomic oxygen on the surface of cadmium selenide film creates electron capture levels, located about 1.2 eV below the conduction band with a concentration ≈ 10^{17} m^{-2} and the electron capture cross section $(8 \div 13) \times 10^{-19}$ m^{-2}.

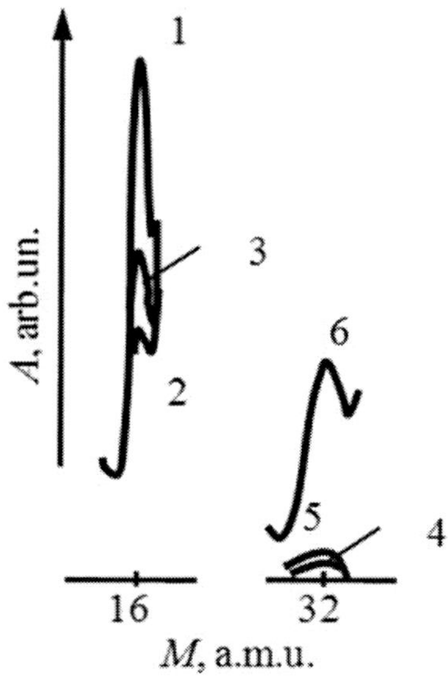

Figure 4.11. The mass-spectrum of atomic oxygen (1-3) and molecular oxygen (4-6) desorption from the surface of cadmium selenide films. 1,4 - initial undoped surface, 2,5 - cadmium-doped surface, 3,6 - indium-doped surface.

Atoms and oxygen molecules have certain selectivity for the certain nature of chemisorption centers on the surface of *CdSe, CdS* semiconductors [1, 53].

The increase in the concentration of excess metal atoms from during surface alloying, mainly *Cd* or *In* (better *In*), leads to the fact that a growing number of chemisorbed molecules O_2 (Figure 4.11). However, surface depletion of the metal atoms leads to the predominance of the atomic form of oxygen chemisorption (Figure 4.12)

E_{bind} O1s=531,8 eV (O_2^- , chem), E_{bind}O1s=531,2 eV (O^- , chem).

It has been found [1, 53] that on the alloyed cadmium selenide surface the binding energy of chemisorbed oxygen molecules with an adsorption center is about 0.75 eV. Some increase in the binding energy of chemisorbed oxygen molecules with centers of its chemisorption on the surface of cadmium selenide films is observed at the initial stage of doping with indium, when the metal is still in an atomically dispersed form.

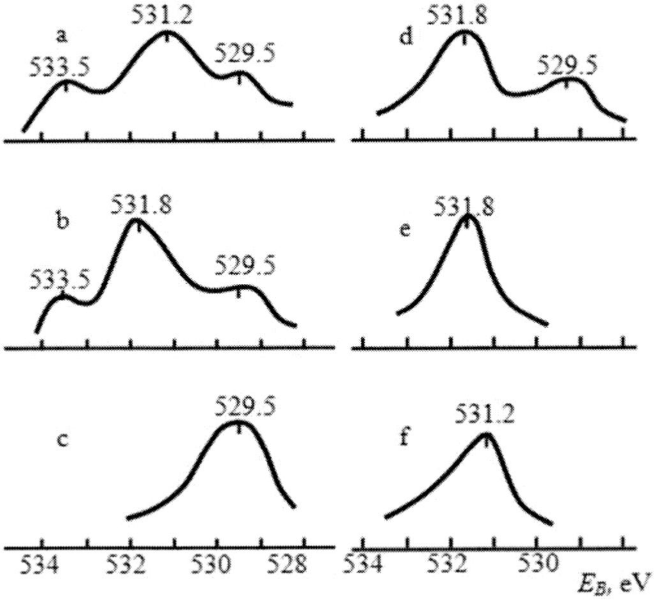

Figure 4.12. *X*-ray photoelectron spectra of *O1s* on the surface of cadmium selenide films: a - initial film of cadmium selenide; b - initial film of cadmium selenide with an excess of cadmium on the surface; c - a film of cadmium selenide after annealing in oxygen atmosphere at $T = 570\ K$ for 20 min; d - cadmium selenide film after prolonged exposure (> 50 hours) in an ultrahigh vacuum $10^{-7}\ Pa$; e - cadmium selenide film, rectified by etching it with Ar^+ ions and with subsequent reaction with dry oxygen with an excess of cadmium on the surface; f - the same as in (d), but the film is derived from the usual charge without enrichment.

In this case, within the resolution of the measurement technique even *In* nanoclusters can not be found. The appearance of the cluster structure of the ligand on the surface with a size of several nanometers or more leads to a reduction of the binding energy of chemisorbed oxygen molecules on *CdSe* sensor surface to 0.66 *eV* with increasing surface concentration of *In* (Figure 4.13, curve 1). This E_{bind} decrease in dipole $In^{\delta+}\ O^{\delta-}$ is due to the fact that a molecule of oxygen after nanocluster formation localizes electron, already owned to *In* cluster structure, and not an isolated In atom on the sensor surface. *In* addition, as the concentration of such dipoles rises, E_{bind} may decrease as a result of the dipoles interaction with each other. The observed E_{bind} variance with the *In* concentration changing is associated with the variance of cluster sizes along the surface (Figure 4.14), which, in turn, reflects the inhomogeneity of the original surface of CdSe sensor.

The decrease of binding energy of the chemisorbed molecules with the surface of the sensor as a result of *CdSe* doping by indium, according to the theory [56], should lead to a decrease in the sensor adsorption sensitivity β. The increment of β with duration of *CdSe* surface doping by *In* atoms up to (40-50) % filling it with metal clusters was experimentally observed (Figure 4.15). This is due to the fact that the influence of Y_s decrease and adsorption centers concentration increase is predominant. Further doping of the surface leads to a decrease of β because of shorting the sensor surface by formed metal structure.

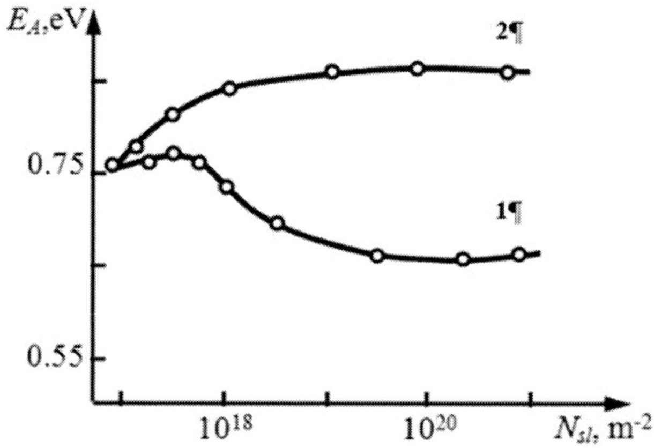

Figure 4.13. The dependence of the binding energy E_{bind} of oxygen molecules with the surface of cadmium selenide films on the degree of doping by atoms of *In* (1) and *Se* (2).

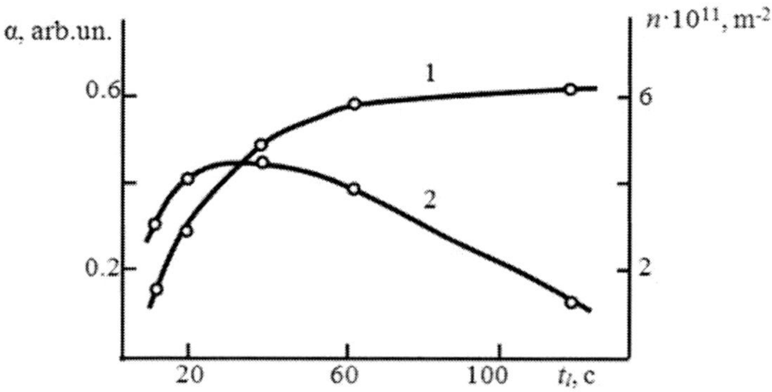

Figure 4.14. The dependence of the surface coverage α (curve 1) by indium clusters and their concentration n (curve 2) on the surface doping duration.

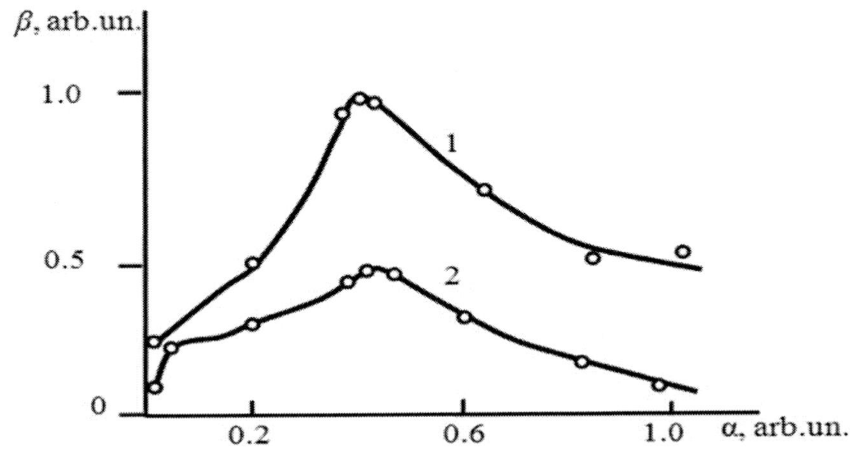

Figure 4.15. The dependence of the adsorption sensitivity β on the relative surface *In* coverage α of *CdSe* films with 1 μm (1) and 5 μm (2) thickness.

In case of reducing the concentration of chemisorption centers of oxygen molecules by surface doping with selenium atoms, the binding energy of O_2 chemisorption centers increases to 0.84 *eV* (Figure 4.13, curve 2), describing the interaction of chemisorbed oxygen molecule with the center of its adsorption as the interaction in isolated complex (electronegativity of oxygen molecule is 0.9 *eV*).

4.4. Effect of Sign Change of Gases Chemisorption on the Properties of Semiconductor Sensor Depending on Its Surface Composition and on the Concentration of Chemisorbed Molecules

Under certain experimental conditions during the interaction of chemisorbed gas particles of the same type on the fixed sensor surface, the sign inversion of the surface chemical reaction is observed. For example, the oxidation effect of chemisorption of certain type of gas particles changes with time to reduction effect.

This phenomenon manifests itself in several variations with a constant gas - sensor reaction pair: a) the change in the type of chemisorption interaction from oxidative to reducing or vice versa, during increasing of concentration of gas in the ambient atmosphere and, consequently, on the surface of the sensor at a constant temperature and sensor surface composition; b) the sign inversion of gas chemisorption due to the influence of changing the composition of sensor surface at constant interaction temperature and concentration of chemisorbed gas in the atmosphere; c) the changes in chemisorption interaction from oxidative to reductive (or vice versa), stimulated by the temperature at constant concentration of chemisorbed gas and surface composition.

Together or separately, these effects were observed in several studies [1, 39, 42, 58-71] for the interaction of CO, SO_2, H_2S, H_2, CCl_4, O_2 with the surface of CdS, $CdSe$, SnO_2, $SnWO_4$, Fe_2O_3, ZnO, Pt/TiO_2.

In the case of CdS films with close to the stoichiometric surface composition, the chemisorption of CO in binary gas mixture (CO, N_2) with its constant concentration shows only donor impact on the sensor conductivity (Figure 4.16, curve 1) [1, 39, 42]. Increase in the concentration of cadmium in CdS sensor produces effect of acceptor chemisorption influence of CO adsorption on the sensor conductivity in the region of relatively low temperatures, $T < 320\ K$ (curve 2). Further cadmium doping of sensor brings to the sensor sensitivity to the CO chemisorption of acceptor type already at room temperatures and, the temperature of chemisorption type changing from acceptor to donor decreases to (290-300) K.

In case of Sn_xWO_y, the increase of tin impurities in the sensor leads to the appearance of the dual effect of CO chemisorption influence with temperature increasing of the acceptor ($T > 230\ K$) to the donor ($T > 250\ K$) at a constant CO concentration value of 250 ppm in dry nitrogen [1, 39, 42]. In the case of sensors, containing a small concentration of excess Sn atoms, and showing the donor response to the chemisorption of 10 *ppm* CO in dry N_2 at $T = 473\ K$, we can make a change to the acceptor-type response by increasing of CO concentration up to values higher than 50 *ppm* (Figure 4.17).

Increasing the number of metal atoms in ZnO [66] and Fe_2O_3 [67] films leads to a change in the charge of chemisorbed CO molecules on film surface from positive to negative. Depending on the partial pressure of CO molecules, they are chemisorbed on Pt/TiO_2 as $CO^{\delta-}$ at low concentrations, while concentration increasing changes the chemisorption complex charge to $CO^{\delta+}$ [59].

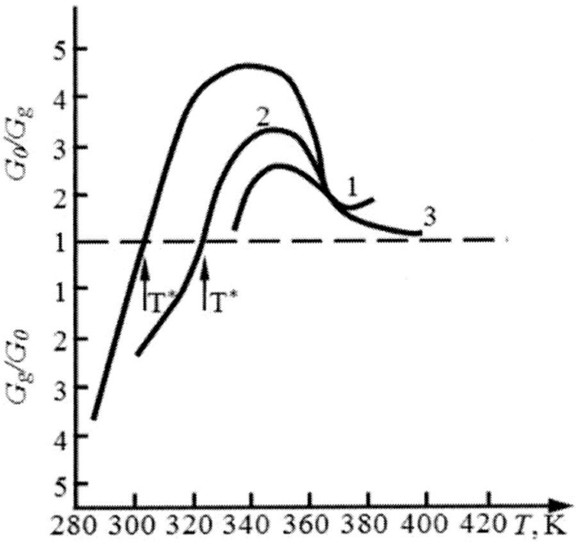

Figure 4.16. Relative conductivity *Gg/Go* (and *Go/Gg*) of cadmium sulfide films with cadmium amount increasing, respectively I, II and III type at 110 *ppm CO* in nitrogen, depending on the temperature. 1 - *CdS* I type, 2 - *CdS* II type, 3 - *CdS* III type *T** - transition temperature from oxidation to reduction behavior of *CO*.

In the case of *CdS* films with composition close to stoichiometric, chemisorption of *CO* molecules on the surface in the form of carbonyl absorption 2000-2300 cm^{-1} results to the appearance clearly defined peak at 2140 cm^{-1} in the IR spectra [39], the intensity of which increases with temperature growth from 290 *K* to 340 *K* (Figure 4.18, a).

Significant doping of *CdS* sensors by excess *Cd* atoms leads to the appearance, except the peak 2140 cm^{-1}, two more maxima in the infrared spectrum localized at 2110 cm^{-1} and 2090 cm^{-1} (Figure 4.18, b) after surface chemisorption of *CO* from the gas mixture with dry N_2 (100 *ppm CO*). At the same time acceptor component of *CO* chemisorption also appears, and its impact depends on the temperature of the sensor (Figure 4.16) and the concentration of *CO* in the gas molecule (Figure 4.17). During the evacuation of *CO* from the reaction chamber, the 2110 cm^{-1} band disappears at first, and then, 15 minutes later, the 2090 cm^{-1} band in IR spectrum disappears.

In *CdS* the role of adsorption centers can play either cadmium atoms, which are located at the sulfur vacancy, or sulfur vacancy itself. In this case, the chemisorbed molecule acts as a donor and it is manifested as 2140 cm^{-1} maximum in the IR spectrum.

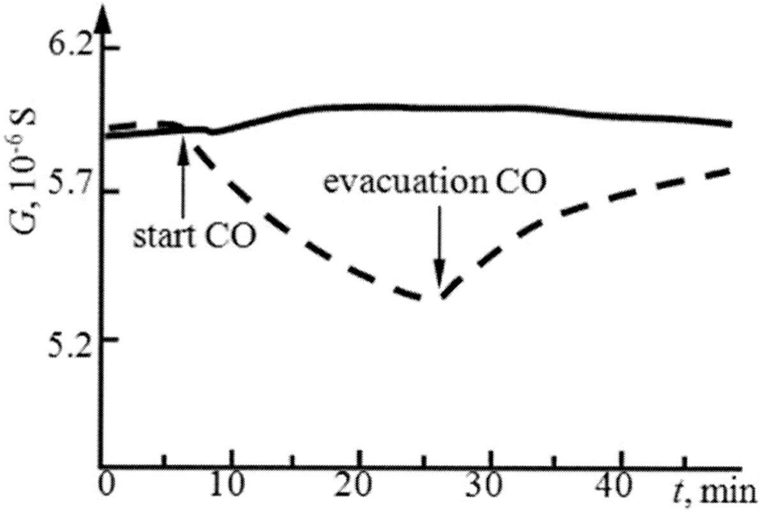

Figure 4.17. The kinetics of the conductivity of Sn_xWO_y films, doped by *Sn* atoms, at different concentrations of *CO* in synthetic dry atmosphere (N_2, *CO*) at 470 *K* (10 ppm CO, - - - 50 *ppm CO*).

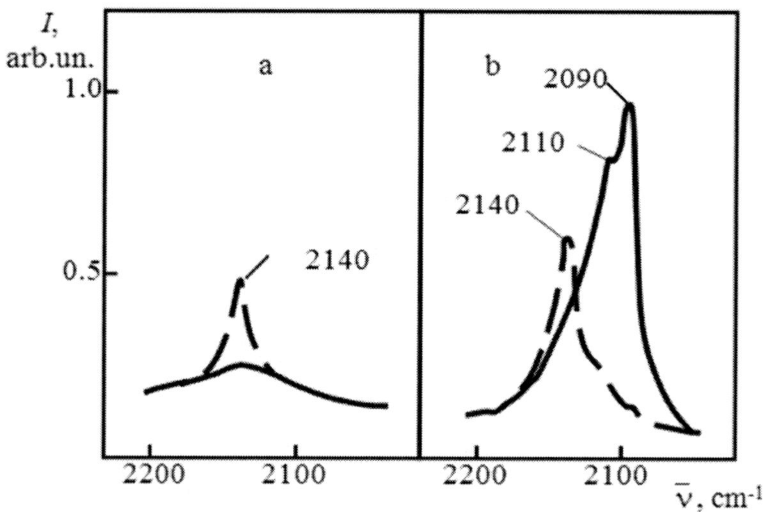

Figure 4.18. IR spectra at carbonyl adsorption region, measured after chemisorption of CO on the surface of *CdS* films of II (a) and III (b) types at 290 *K* (solid line) and 340 *K* (dotted line).

In case of doping by cadmium the variety of chemisorption centers nature may increase due to interstitial atoms Cd_i, *Cd* atoms on the surface in an atomically dispersed form or in its micro and macro clusters. We can assume that in this case the *CO* molecules act as acceptors at chemisorption on *CdS* sensors that manifest themselves as responsible for the IR peaks at 2090 cm^{-1} and 2110 cm^{-1}.

The differences in the kinetics of desorption show that either acceptor interaction of the *CO* molecule on the surface of *CdS* is implemented on two different defects with different binding energy, or by interacting with the same defect, the *CO* molecule, depending on the configuration of communication generates two different types of communication, different structure and performance characteristics.

The results of works [39, 42] show that the phenomena when the nature of the electrodes material changes, changing the semiconductor conductivity type (wide-band materials are used) or the changes in the composition of sensors during heating (used relatively low T <500 *K*) cannot be responsible for the observed dual effect of gases chemisorption on the sensors properties.

Carbon monoxide in the carbonyl form at relatively low T is coordinated on the surface metal atom via carbon atom [68]. On the other hand, apart from the linear structure of the binding of the *CO* molecule to the surface [68, 72], in some cases, as well as basic can be viewed and bridge structures [72], which are formed by a pair of 1π - *CO* orbitals and empty orbitals of the metal or σ-bonds in the case of sp^2 - hybridization of carbon atom. IR absorption bands below 2000 cm^{-1} are associated with bridge structures, and IR absorption bands located in the region of 2000 cm^{-1} – with a linear structure.

In this case, as the most likely a linear structure is considered [70], suggesting that the formation of bridged configurations requires some specific conditions. Thus, in the case of iron oxide and chemisorbed *CO* molecule with a positive and a negative charge forms a linear structure, the charge states of which depend on the stoichiometry of the sensor surface [73].

On the surface of metal oxides *CO* molecule can be chemisorbed or by the metal atoms or by oxygen atoms, also depending on sensor stoichiometry. In this case, due to the characteristics of the electric structure, *CO* molecules can form various adsorption types, which will differ in structure and orientation relative to the sensor surface, in the charge state and dipole moment. The implementation of the donor or acceptor types of carbon monoxide chemisorption is also depending on the nature of sensor material, on condition of its surface, the extent of its coverage and on the terms of interaction (T, pressure, composition of the gas atmosphere, and so on).

4.5. Anomalous Donor Effect of Oxygen Chemisorption on the Electrical Conductivity of *N*-Type Cadmium Selenide Films

The found effects of the ligand atoms influence on form of gas particles chemisorption, their binding energy with the centres of chemisorption on the sensor surface, in some cases, allow us not only to assume the mechanisms of surface interactions, but also, under certain conditions, to control these processes [1, 40-42, 49 , 74-89]. *CdSe, CdS* films surface alloying by atoms of donor metals, the best of which is *In*, as it follows from the results of *X*-ray photoelectron spectroscopy – *XPS*, secondary-ion mass spectrometry - *SIMS, Auger* spectroscopy, surface potential and concentration of the carrier mobility investigation, and other measurements, leads to a change in the basic form of oxygen chemisorption from the atomic to molecular one. In this case, the chemisorbed molecule chooses excess metal atom as the adsorption centre, which is registered by the corresponding changes in the *X*-ray photoelectron spectra.

The binding energy of the molecules with trapped electrons at the centre of chemisorption is about 0.75 *eV*, and appears as the presence of the local energy level within semiconductor band gap. Oxygen atom creates a local level in the band gap of about 1.2 *eV* from the bottom of the conduction band. As a rule, the emergence of a chemisorbed oxygen atom is the result of the dissociation of chemisorbed O_2^- molecules, which, due to captured negative charge, interatomic bonds are weakened.

Doping of the surface by atoms of the acceptor impurity, which *Se* atoms are, leads to the weakening influence of the acceptor oxygen chemisorption with *Se* concentration increasing (Figure 4.9) due to the reduction of the probability of electron hitting to the surface from the conduction band due to the increase of the surface potential barrier. After *p*-type conductivity layer forming on the sensor surface in the high N_{Se} region (Figure 4.9) [1, 39, 49], the chemisorption of oxygen, which, as before, captures the free electrons of the sensor, leads, naturally, to the growth of *p*-type conductivity. This effect, without proper analysis, can be seen as an apparent donor influence of acceptor oxygen gas chemisorption. A similar phenomenon may occur in the case if, due to any reason, on the surface of n-type sensor a fine channel of *p*-type conductivity appears, which interacts with chemisorbed acceptor particles and determining the sensor conductivity. Therefore, the discussion of the donor acceptor particles chemisorption effect on the surface of *n*-type conductivity sensors can be made only on the basis of detailed knowledge of the properties and composition of the surface.

"Donor" influence of oxygen chemisorption on the electrical conductivity is shown in the case of sensors with n-type conductivity with the indium atoms doped surface - *CdSe: In* (Figure 4.19, curve 2). These are the sensors that have the effect of acceptor chemisorption, as shown in the previous sections, should be most pronounced.

It was found [52] that the anomalous effect of donor oxygen chemisorption is expressed as that at room temperature oxygen chemisorption of dried gas mixture (O_2, N_2) in the initial stages of chemisorption of O_2 accompanied by an increase of electrical conductivity, which is then followed by the usual decrease in current sensor. Time τ_{max} of maximum achievement in the kinetics curve $I(t)/I_{max}$ depends on temperature and decreases with the speed of filling the measuring chamber with a gas mixture, and the height of the maximum increases. τ_{max} decreases exponentially with the activation energy (0,35 ± 0,02) eV with temperature increasing (Figure 4.20). The intensity of the kinetics maximum on $I(t)/I_{max}$ curve increases with increasing of doping indium atoms concentration on the surface (Figure 4.21), but decreases with sensor temperature growth (Figure 4.21, curves 1 – 4).

The foregoing facts show that the appearance of "donor" effect of oxygen chemisorption on *CdSe:In* sensor occurs as the result of the simultaneous development of two opposing processes that differ in their time constants.

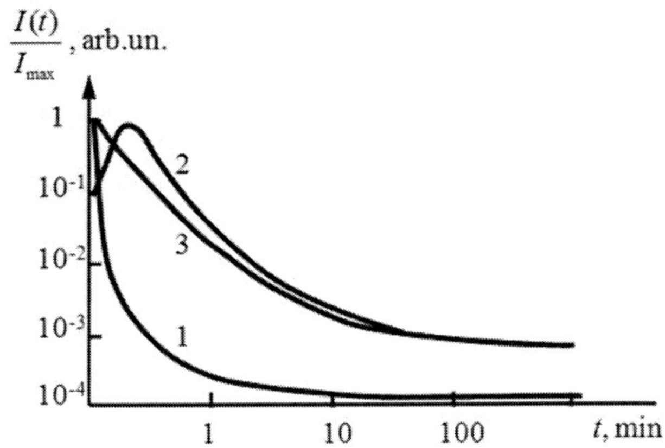

Figure 4.19. The kinetics of the dark current with the rapid oxygen concentration changes in the test chamber, in which *CdSe* (1), *CdSe: In* (2), sensors were put, and the same with slow change in the concentration of oxygen at the surface - *CdSe: In* (3) sensors.

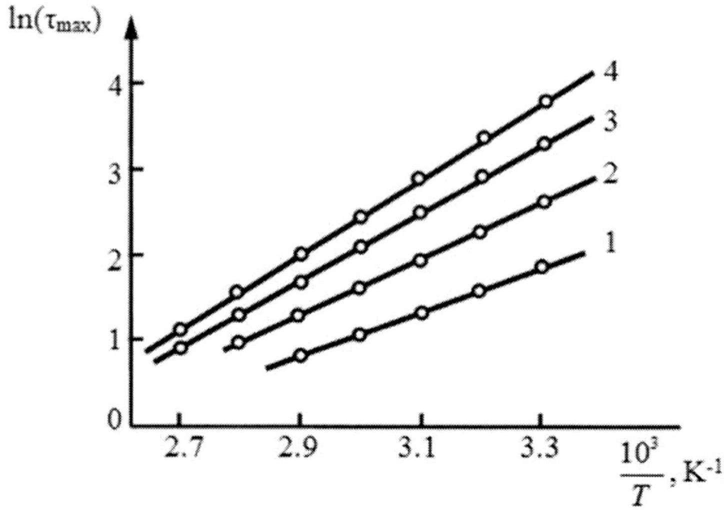

Figure 4.20. Temperature dependence of τ_{max} on the curve of the dark current kinetics for oxygen chemisorption on the surface of the *CdSe:In* sensor with the doping degrees of *In*: $10^{17}\ m^{-2}$ (1), $5\times10^{17}\ m^{-2}$ (2), $10^{18}\ m^{-2}$ (3) $5\times10^{18}\ m^{-2}$ (4).

The first of these processes - donor conditioning effect, grows rapidly with an activation energy ≈ 0,35 *eV*. The second process – acceptor one, implemented more slowly and causes kinetics features on its later stages.

Observed "donor" effect of chemisorption of acceptor-type gas particles on the surface of n-type semiconductor sensor is caused by recharging of the impurity-origin levels, which are only about *kT* below the Fermi level on the surface [52] in a narrow energy range. If the concentration of these centers in the sensor original state after alloying before chemisorption, is comparable to the concentration of free electrons in the conduction band at the surface, the result of the process will carry on in the following line. Intake of chemisorbed oxygen on the surface leads to the accumulation of negative charge from the conduction band on the surface oxygen levels, which entails an increase in the surface potential barrier and caused by this removal of the Fermi level from the bottom of the conduction band.

As a result, the position of the Fermi level crosses the energy levels of impurities, the electrons are emitted from them in the conduction band and cause the "donor" growth of $I(t)/I_{max}$. In this case, the higher the temperature, the closer Fermi level is to the levels of impurities and the earlier τ_{max} is achieved. At a sufficient temperature impurity levels may already get empty, and then only the acceptor effect of oxygen chemisorption is observed.

If the levels, that situated about *kT* below Fermi level, would be caused by own sensor surface defects, the height of the maximum dependence on concentration of indium dopant atoms would not be observed. However, as indicated by Figure 4.21, the I_{max}/I_0 ratio dependence on N^s_{In}, that indicates the surface nature of observed "donor" phenomenon.

The data in Figure 4.19, 4.20 and 4.21 show that the actually oxygen interacts with the surface of n-type *CdSe:In* sensor as an acceptor particle, and the "donor" effect arises from the surface charge of the impurity in the level directly below the Fermi level, stimulated by an increase in the surface potential barrier due to accumulation of negative charge at the levels of the chemisorption origin.

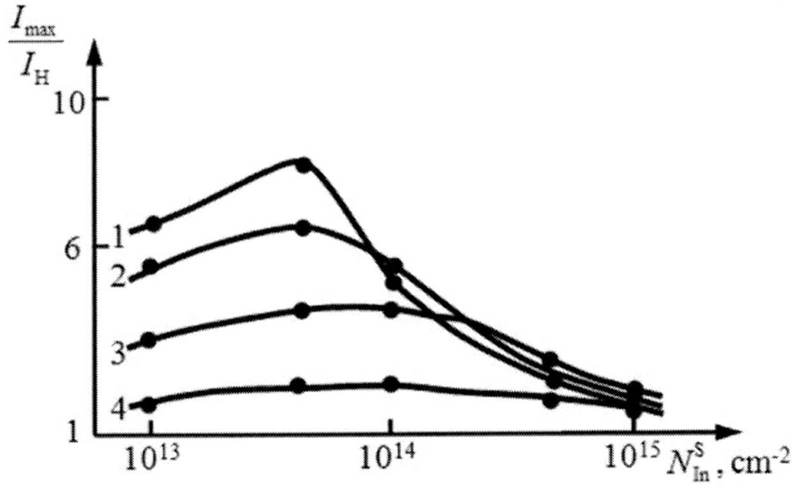

Figure 4.21. I_{max}/I_H depending on the degree of indium doping of cadmium selenide surface, measured at 293 *K* (1), 413 *K* (2), 433 *K* (3), 453 *K* (4), I_H - current value before doping.

The possibility of observing this process depends on the rate of oxygen supply to the sensor surface, its concentration, the dopant concentration and the position of its energy level relative to the Fermi level, temperature sensor and, therefore, it is not so much as an anomaly, but as a rare because of the complexity of simultaneous satisfaction of the conditions and location in the required range of all of these parameters.

4.6. The Effect of *N*-Type Sensor Surface Alloying on the Chemisorption Kinetics of Acceptor Oxygen Particles

As it's shown in this chapter, indium is the most effective ligand for the controlling of chemisorption properties of the sensor surface of *n*-type *CdSe, CdS*. Comparing the mass spectra in Figure 4.11. one can see that by increasing the concentration of alloying indium atoms it is possible to change the basic type of oxygen chemisorption from atomic to molecular. In this case, indium condenses in an atomically dispersed form, as evidenced by the observed *X*-ray photoelectron spectra, characteristic for free *In* atoms (Figure 4.22), and forms a micro-macroclusters, that creates the grid and other structures on the sensor surface (Figure . 4.6).

Increasing the concentration of *In* atoms on the surface of the *CdSe* sensor up to 10^{18} m^{-2} stimulates an increase in its electrical conductivity (Figure 4.23, curve 1, 2) due to increase in the concentration of shallow, ionized at room temperature, surface donors, which are indium atoms [1].

The positive charge of these ionized *In* atoms compensates the biographical negative charge accumulated on the sensor surface after manufacturing before alloying process.

This entails a reduction in the height of the surface potential barrier, and, therefore, brings to increase of the probability of a free electron coming to the surface.

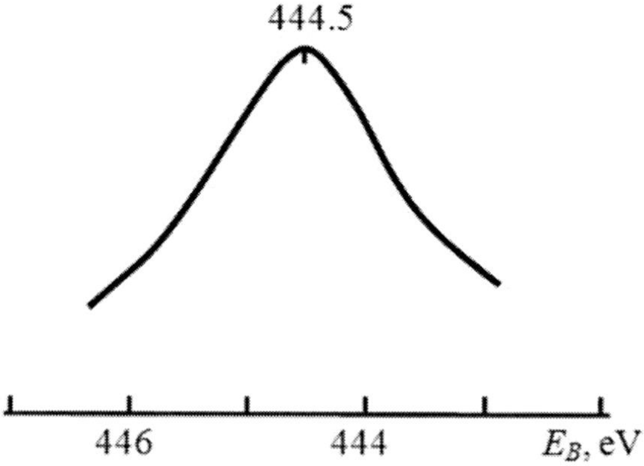

Figure 4.22. *X*-ray photoelectron spectrum of *In3d5/2* on the cadmium selenide film surface.

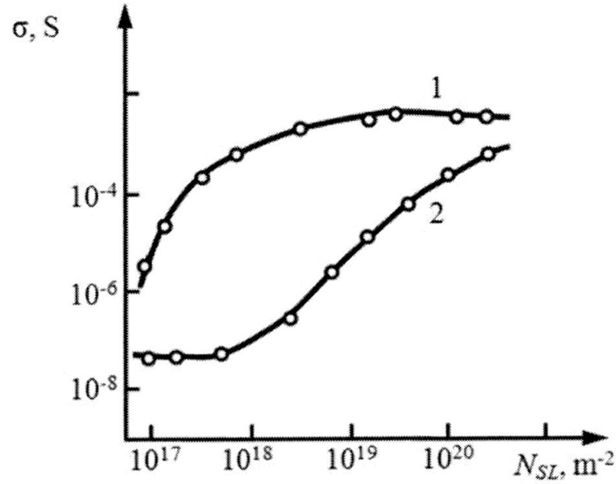

Figure 4.23. Dependence of *CdSe* sensors conductivity on duration of their surface doping by indium atoms, measured in vacuum before exposure to oxygen (1) and after exposure to O_2 atmosphere (2).

This fact, as well as increasing the concentration of oxygen chemisorption centers, which are indium atoms deposited on the surface of the sensor, together cause an increase in the sensitivity of the sensor surface adsorption as a result of doping up to concentrations of indium atoms of the order of 10^{18} m^{-2} (Figure 4.7).

With increasing of doping time the surface coverage by dispersed atoms of indium and its micro and macroclusters also increases (Figure 4.14, curve 1). This increment is accompanied by increased β up to 40-50% indium filling of the sensor surface.

Continued surface doping by indium brings about the situation in which formed metal micro- and macrostructures on the surface begin to short-circuit the sensor and its adsorption sensitivity decreases.

Processes occurring on the surface as a result of the doping concentration of indium atoms up to 10^{18} m^{-2} lead to a significant reduction of oxygen adsorption τ_a and desorption τ_d times (Figure 4.24).

The observed decrease of τ_a occurs due to the increase in the electron exchange rate, caused by a decrease of φ_s and lowering binding energy of chemisorbed oxygen molecules with an adsorption center from 0.8 *eV* to 0.65 *eV*, as shown in [1]. At the same time, the decrease of desorption time τ_d

occurs only as a result of reducing the binding energy of the adsorbate to the sensor surface.

Chemisorption interaction of oxygen with semiconductor sensor surface is a kind of chemical reaction, the dependence of the constants of which on the temperature is described by the Arrhenius equation [90]

$$K = K_0 \exp\left(-\frac{\Delta W}{KT}\right)$$

(K_0 - reaction constant at $0\ {}^0K$, ΔW - activation energy). The value of the adsorption time depends on measurements temperature, as it can be seen from Figure 4.24.

And the rate of adsorption decrease of sensor conductivity (Figure 4.25), as the rate of its desorption increase of conductivity (Figure 4.26) is exponentially dependent on temperature.

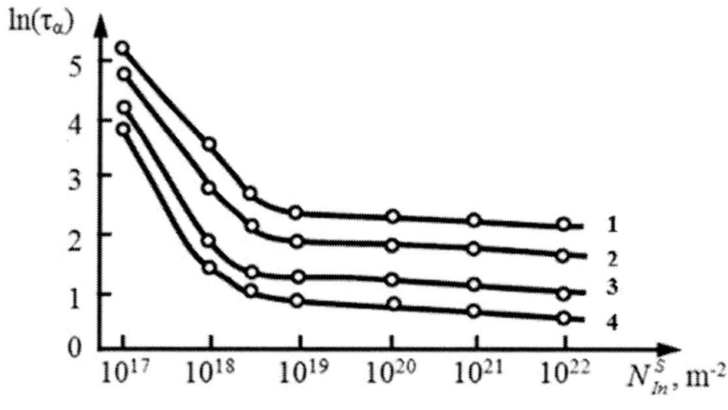

Figure 4.24. The dependence of the adsorption time τ_a for oxygen chemisorption on cadmium selenide films on the degree of surface doping by indium atoms. Measuring temperature: 343 K (1), 453 K (2), 363 K (3), 373 K (4).

Processing of the results from Figure 4.25 gives the activation energy value of (0,35 ÷ 0,4) eV. Desorption is controlled by removal process of electrons to the conduction band from the levels of chemisorbed oxygen molecules, which, as follows from Figure 4.26, requires the activation energy (0,7 ÷ 0,8) eV. So the intensity of oxygen desorption is determined by the rate of chemisorption bond breaking and release its trapped electron in the conduction band of the sensor.

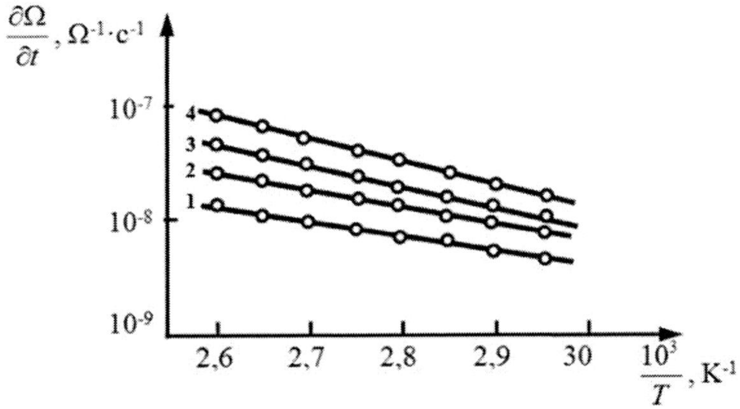

Figure 4.25. Temperature dependence of the rate of decrease in the electrical conductivity during oxygen chemisorption on *CdSe:In* films (1), (N^s_{In} = 10^{17} m^{-2}); *CdSe:In* (2), (N^s_{In} =10^{18} m^{-2}); *CdSe:In* (3), (N^s_{In} =10^{19} m^{-2}); *CdSe:In* (4), (N^s_{In} =10^{20} m^{-2}).

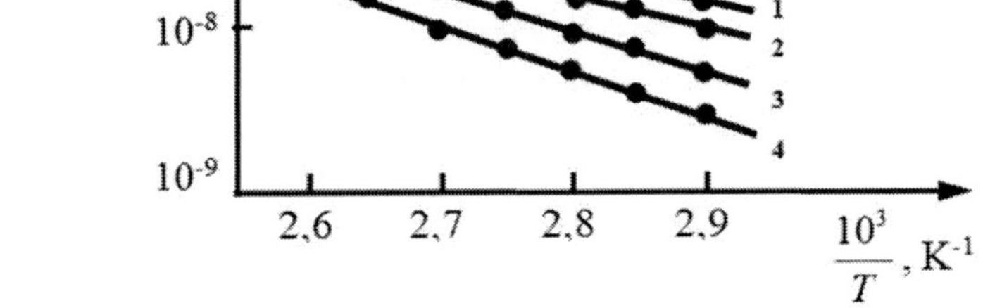

Figure 4.26. The temperature dependence of the rate of increase in the electrical conductivity during the oxygen desorption from the surface of cadmium selenide films *CdSe:In* (1) (N^s_{In} =10^{17} m^{-2}); *CdSe:In* (2) (N^s_{In} =10^{18} m^{-2}); *CdSe:In* (3) (N^s_{In} =10^{19} m^{-2}); *CdSe:In* (4) (N^s_{In} =10^{20} m^{-2}).

Activation energy of 0.35 *eV*, obtained in the study of oxygen chemisorption (Figure 4.25, curves 1-3), can not be associated with the presence of a potential barrier of electrostatic origin, since an increase in the donor concentration on the surface of an *n*-type, the surface is enriched by electrons and the reduction of the surface potential barrier occurs. For $N_{In}^{s} >$ 5×10^{18} m^{-2}, according to the views discussed in [1, 56], indium atoms are adsorbed in the neutral form and the electrical layer is close to degeneration. Therefore, the Fermi level should be near the bottom of the conduction band and, thus, on the surface there is no electrostatic barrier for the movement of electrons from the bulk to semiconductor surface. Consequently, the energy of 0.35 *eV* is connected with the existence of the barrier, having a chemical nature, and can be treated as a threshold needed to move a molecule of oxygen from the neutral form to the charged one. This chemical nature barrier may be responsible for the 0.35 *eV* activation energy of "abnormal donor" effect kinetic of oxygen chemisorption.

REFERENCES

[1] Smyntyna, V. Electron and Molecular Phenomena on the Surface of Semiconductors. New York: NOVA Publishers; 2013.

[2] Снитко, ОВ. Исследование электрофизических свойств поверхности *Ge, Si, CdS* при различных активных воздействиях: *Дис. ... д-ра физ.-мат. наук* .К; 1968, 393 с.

[3] Мясников, ИА. Об особенностях адсорбции молекул кислорода и радикалов на поверхности окислов металлов в газовых и жидких средах. *Журнал физической химии*, 1981 т. 55, 1278–1282.

[4] Карпович, ИА. Кинетика изменения потенциала поверхности пленок CdSe при поверхностном легировании. В: *Физика полупроводников и полупроводниковая электроника.* Саратов: изд-во Саратовского университета; 1970 вып.5, 3–10.

[5] Karpovich, IA; Kalinin, AN. Influence of surface doping on the surface potential and surface conductivity of cadmium selenide films. *Soviet physics. Semiconductors*, 1970. — vol.12, no. 5, 1490–1495.

[6] Brillson, LJ. Chemical reactions and local redistributions at metal-CdS and CdSe interfaces. *Physical Review*, 1978 vol. B18, no. 6, 2431–2446.

[7] Миленин, ВВ; Примаченко, ВЕ; Снитко, ОВ Влияние примеси палладия на электронные свойства поверхности кремния. *Украинский физический журнал*, 1979 т. 24,. 910–915.

[8] Бедный, БИ; Калинин, ЛН; Карпович, ИА. О стабилизации эффекта поверхностного легирования пленок селенида кадмия. *Микроэлектроника*, 1976 т.5, вып. 3, 289–290.

[9] Brillson, LJ. Metal-induced surface states on CdS and CdSe. In: *Phys. Semicond.Proc.13-th Int. Conf.*, Rome., 1976, 665–668.

[10] Снітко, ОВ; Міленін, ВВ; Примаченко, ВС. Електронні властивості поверхні кремнію при сумісній адсорбції домішок золота і срібла. *Доповіді Академії наук Української РСР. Сер. "А", фізико-математичні науки*, 1975 вип.4, 372–375.

[11] Белановский, А С. Адсорбция и десорбция на поверхности полупроводников. В: *Обзоры по электронной технике. Сер. Микроэлектроника*. 1969 вып. 58, 126.

[12] Karpovich, IA; Kalinin, AN; Bedniy, BI. Definition of parameters of the semiconductor surface with surface doping. *Soviet physics. Semiconductors*, 1976 vol. 10, no.7, 1402–1407.

[13] Мясников, ИА; Большун, ЕВ; Агаян, ВО. Об ионизации атомов металлов на пленках окиси цинка и о полупроводниковых детекторах. *Доклады АН СССР*, 1975 т. 220, 1120.

[14] Voorhoeve, RJH; Jonson JrDW; [and others]. Perovskite oxides: material science in catalysis ($CaTiO_2$)-ABO_2. *Science*, 1977 vol. 195, no. 428, 827–833.

[15] Brillson, LJ. Surface electronic and chemical structure of (1120) CdSe: comparison with CdS. *Surface Science*, 1977 vol. 69, 62–84.

[16] Коробко,ГС;. Вашпанов, ЮА; Карлов,ОГ; Бойчук, ВН. Чувствительный элемент газоанализатора на кислород. *Авторское свидетельство. № 1514079 СССР*, МКИ G01N27/02ю № 3359062; Заявлено 06.11.88; Опубл. 08.06.89.

[17] Bedniy, BI; Ershov, SN; Kruglov, EN. Effect of surface doping of gallium arsenide on the characteristics of gold contacts Al — (h^+ — GaAs*). Soviet Physics Journal*, 1985 no.2, 80–83.

[18] Davydov, MA; Kiselev, SA; Palatnik, GM; Ruzin, VA; Sokolova, NP. Studies nature of the surface of highly dispersed palladium adsorbents. *Physics, Chemistry and Mechanics of Surfaces*, 1984 no.4, 84–90.

[19] Arutyunyan, RV; Baranov, VYu; Bolshov, LA; Borzenyuk, VL. Doping of solid surfaces from a plasma laser breakdown in liquids. *Physics, Chemistry and Mechanics of Surfaces*, 1984 no.4, 149–151.

[20] Andriyahin, VN; Vasil'ev, VY; Edieral, NV; Kletskin, YaG. About passivation of the surface of iron-based alloys after laser alloying with chromium. *Physics, Chemistry and Mechanics of Surfaces*, 1984 no.4, 145–148.

[21] Романов, ОВ. Закономерности и особенности воздействия адсорбции микропримесей металлов на электрофизические свойства Ge и Si. В: *Препринт* № 29–78. Новосибирск: 1978.

[22] Milenin, VV; Primachenko, VE; Snitko, OV. Influence of impurities on the electronic properties of palladium-silicon surface. *Ukrainian Journal of Physics*, 1979 vol. 24, no. 7, 910–915.

[23] Мясников, ИА; Поспелова, ИН; Карецкая, ТА. О структуре и электропроводности поликристаллических пленок окисных полупроводников. *Доклады АН СССР*, 1968 т.179, 695.

[24] Миленин, ВВ; Полудин, ВИ; Примаченко, ВВ; [и др.]. Образование дисперсных структур Ag и Au при их адсорбции на поверхности Si. *Известия Академии Наук СССР. Неорганические материалы*,1977 т.13, 517–515.

[25] Morrison, SR. *The Chemical Physics of Surfaces*. New-York and London: Plenum Press; 1977.

[26] Миленин, ВВ; Примаченко, ВЕ; [и др.]. Исследование электронных свойств поверхности кремния, легированной металлами с электроотрицательным электрохимическим потенциалом. *Физика и техника полупроводников*, 1979 vol.13, вып. 8, 1532–1538.

[27] Волькенштейн, ФФ. Электронные уровни атомов, адсорбированных на поверхности кристалла. *Журнал физической химии*, 1947 т.21, 1317–1321.

[28] Rowe, TF;. Margaritondo, G; Christman, BB. Mechanism of Shottky barrier formation. *Physical Review*, 1978 vol. B15, 2195–2198.

[29] Brillson, LJ. Chemical reactions and local redistributions at metal-CdS and CdSe interfaces. *Physical Review*, 1978 vol. B18, no.6, 2431–2446.

[30] Tatabadze, DG. The effect of adsorption of metal atoms on the electrical conductivity of films of cadmium sulfide. *Russian Journal of Physical Chemistry A*, 1976 vol.50, no.4, 995–996.

[31] Brillson, LJ. Surface electronic and chemical structure of (1120) CdSe : comparison with CdS .*Surface Science*, 1977 vol. 69,62–84.

[32] Бедный, БИ. Исследование электрофизических свойств реальной поверхности некоторых полупроводников при поверхностном легировании металлами. *Автореф. дис. канд. физ.-мат. наук*. Горький: 1977.

[33] Бедный, БИ. Исследование электрофизических свойств реальной поверхности некоторых полупроводников при поверхностном легировании металлами.: *Дис. канд. физ.-мат. наук*. Горький: 1977.

[34] Вашпанов, Ю.А. Адсорбционная чувствительность полупроводниковых материалов группы A_2B_6, оксидов тяжелых металлов и кремния с реальной поверхностью с кластерными структурами.: *Дис. ... докт. физ.-мат. наук*. Научный консультант Смынтына, ВА. Одесса : ОГУ; 1999.

[35] Моррисон, С. *Химическая физика поверхности твердого тела*. М. : Мир;1980.

[36] Golovanov, V: Smyntyna, V; Interaction Between Collective and Local Subsystems in Semiconductor Surface-Active Structures. *Sensors and Actuators B*, 1994 vol. 25, no.1–3, 647–652.

[37] Снитко, ОВ. Исследование электрофизических свойств поверхности германия, кремния, сульфида кадмия при различных активных воздействиях. *Дис. д-ра физ.-мат. наук*. К: КГУ; 1969.

[38] Сминтина, ВА. Науковий керівник. Дослідження електронно-молекулярних явищ на поверхні напівпровідникових шарів і розробка наукових основ фізико-хімічних процесів у напівпровідникових сенсорах. : *Звіт про НДР (Заключний*). № ДР 0100U002859. Одес. нац. ун-т: Одеса; 2002.

[39] Вашпанов, ЮА. Чувствительность электропроводности полупроводниковых пленок к газам. *Деп. ВИНИТИ*, № 2469-Ук88, 1988..

[40] Герасютенко, ВА. Формирование свойств и состава пленок CdSe и CdS в процессе их роста и легирования. *Дис. ... канд. физ.- мат. наук*, научный руководитель Смынтына., ВА. Одесса: ОГУ; 1990..

[41] Смынтына, ВА. Электрофизические, фотоэлектрические и электронно-молекулярные явления, стимулированные хемосорбционно-диффузионными процессами на границах полупроводниковых поликристаллических пленок селенида и сульфида кадмия. *Дис. ... д-ра физ.-мат. наук*. 01.04.10. “Физика полупроводников и диэлектриков”, Институт полупроводников АН УССР: Одесса; 1988 .

[42] Голованов, ВВ. Характеризация полупроводниковых поверхностно активных веществ $Sn_xW_yO_z$, $Cd_xS(Se)_{1-x}$, In_2O_3: *Дис. ... д-ра физ.-мат. наук* : 01.04.10. “Физика полупроводников и диэлектриков” Научный консультант Смынтына; ВА ОГУ: Одесса; 1998.

[43] Baran, SV; Branitsky, GA.; and Ivanovskaya, MI. Thermocatalytic sensors with Pd-Pt-AL_2O_3 catalyst. *Sensors and Actuators B*, 1993 vol. 13–14, 244–247.

[44] Orlik, DR; Ivanovskaya, MI;. [and others]. Monoelectrode gas sensors based on SnO_2 semiconductor ceramics. *Sensors and Actuators B*, 1993 vol.13–14, 605–607.

[45] Yakusawa, Y; Seki, T; and Jun-ichi Muramatsu, Film-type In_2O_3. Sensors and Actuators. B, 1993 vol.13–14, 613–614.

[46] Matsuura, S; New developments and applications of gas sensors in Japan. *Sensors and Actuators B*, 1993 vol. 13–14, P. 7–11.

[47] Lantto, V; and Mizsei, J. H_2S monitoring as an air pollutant with silver doped SnO_2 thin-film sensors In: *Proc. 3rd Int. Meet. Chemical Sensors*, Cleveland, OH, USA, Sept. 24–26. 1990, 39–43.

[48] Takahata, K. Tin dioxide sensors. Development and applications. *Chemical Sensor Technology*, 1988 vol. 1,. 39–45.

[49] Vaspanov, YuA. Electronic and adsorption properties of the real surface of cadmium selenide films containing indium whiskers. *Technical physics letters*, 1977 vol. 23, no. 8, 580-581.

[50] Brillson, L J. Chemical reactions and local redistributions at metal-CdS and CdSe interfaces. *Physical Review*, 1978 vol. B 18, no.6, 2431– 2446.

[51] Rowe, GE; Margaritondo, G; Christman, L. Mechanism of Scottky barrier formation. *Physical Review*, 1977 vol. B 15, 2195–2198.

[52] Сердюк, ВВ; Смынтына, ВА. Разработка адсорбционных полупроводниковых макетов датчиков концентрации кислорода в газовых средах: *Отчет о НИР (заключительный)* Одесса: Одес. гос. ун-т.; Инв. № 02840010300].: М: ВНИИТИ; 1983

[53] Татабадзе, ДГ. О влиянии адсорбции атомов металлов на электропроводность пленок CdS . *Журнал физической химии*, 1976 т.50, 995–996.

[54] Смынтына, ВА; Герасютенко, ВА; Корнева, СА. Влияние легирования на адсорбционно-десорбционную чувствительность пленок селенида кадмия. *Известия АН СССР. Сер. Неорганические материалы*, 1989 т.29, 856–859.

[55] Bube, Richard H. *Photoconductivity of solids* .New York; London: John Wiley & Sons; 1960.

[56] Вашпанов, ЮА; Смынтына, ВА. *Адсорбционная чувствительность полупроводников*. Одесса: Астропринт; 2005.

[57] Козлов, СИ. О кинетике заряжения поверхности полупроводника при адсорбции. *Известия вузов. Сер. Физика*, 1975 вып. 2, 116–120.

[58] Torvela, H; Pijolat,C; Lalauze, R. Dual response of tin dioxide gas sensors characteristic of gaseous carbon tetrachloride. *Sensors and Actuators B*, 1991 vol.4, 445–450.

[59] Schierbaum, KD; Kirner,UK; Geiger, JF; Gopel, W. Schottky-barrier and conductivity gas sensors based upon Pd/SnO_2 and Pt/TiO_2. *Sensors and Actuators B*, 1991 vol. 4, 87–94.

[60] Moseley, PT; Norris, OW. *Techniques and mechanisms in gas sensing*. Adam Hilger: Bristol, New-York; 1991.

[61] Голованов, В; Гудис, А; Смынтына, В Полупроводниковый сенсор на основе CdS *Журнал аналитической химии*,1991 vol.46, 2374–2379.

[62] Solis, JL; Golovanov, V; Lantto, V; Leppavuori, S. A study of dual conductance response to carbon monooxide of CdS and $SnW0_4$ thin films. *Physica Scripta*, 1994 vol. 54, 248–251.

[63] Lantto, V; Rantala, T. Equilibrium and non-equilibrium conductance response of sintered SnO_2 samples to CO. *Sensors and Actuators B*, 1991 vol. 5, 103–107.

[64] Chiotti, G; Boccuzzi, F; Ciorino, A. Surface characterization of Cu/ZnO catalysts: IR, VIS, UV study of CO chemisorptions. In: *Adsorption and Catalysis on Oxide Surfaces*. Amsterdam: Elsevier Science Publishers B.V.;1985.

[65] Moseley, PT. Materials selection for semiconductor gas sensors. *Sensors and Actuators B*, 1992 vol.6, 149–156.

[66] Ворошилов, ИГ; Роев, МН. Формы адсорбции молекул CO на ZnO. *Доклады Академии Наук УССР*, Б, 1973 Вып. 9, 830.

[67] Ковалев, ГИ; Каган, ЮБ; Крылова, АВ . Заряженные формы хемосорбции окиси углерода на железном катализаторе синтеза спиртов из окиси углерода и водорода. *Кинетика и катализ*, 1969 т. 10, вып. 3, 596–603.

[68] Schroter, Von W; Lautenschlager, KH; Bibrack, H; Schabel, A.. *Chemie*. Leipzic: VEB Fachbuchverlag; 1986.

[69] Kasowski, AV; Rhodin, T; Tsai, M. Chemisorptive bonding of carbon monoxide on nickel (001): formulation and application of a new pseudofunctional electron muffin tin approach. *J. Appl. Phys*., 1986 vol. A41, no. l, 61–72.

[70] Knozinger, H; Che, M; Bond, GC. Surface structure and nature of adsorption sites of supported, reduced and sulfied molybdate catalysts. In: *Adsorption and Catalysis on Oxide Surfaces*. : Amsterdam: Elsevier Science Publishers B. V.; 1985.

[71] Краснов, КС. *Молекулярные постоянные неорганических соединений*. Химия: Л; 1979.

[72] Даньшина, ВВ; Кировская, ИА. Спектроскопические исследования адсорбции малоатомных молекул на поверхности соединении A_2B_6 . В: *Тез. докл. X Всесоюзн. семинар*, 6–8 сент. 1988.

[73] Лебедев, НИ; [и др.]. Формы хемосорбции молекулы CO на поверхности Fe_2O_3 различного стехиометрического состава. *Журнал физической химии*, 1988 т.62, вып. 2, 435-437.

[74] Smyntyna, VA; Moin, MD; Gerasyutenko, VA; Kaganovich, EB; Korneva, SA; Pochtar ,EM. Effect of laser treatment on the adsorption interaction of the films of cadmium sulfide and oxygen. *Soviet Physics Journal*, 1990 vol. 33, 82–85.

[75] Smyntyna, VA; Gerasyutenko, VA; Korneva, SA. Change the surface properties of cadmium selenide films during their growth. *Bulletin of the Academy of Sciences of the USSR. Inorganic Materials*, 1987 vol.23, no.11, 1814–1817.

[76] Smyntyna, VA. Properties of the real surface of cadmium selenide thin films. *Bulletin of the Academy of Sciences of the USSR. Inorganic Materials*, 1982 vol.18, 1263-1266.

[77] Smyntyna, VA; Senkevich, AI; Gerasyutenko, VA; Latypov, SI; Dmitriev, AG. Effect of nature of defects on the electron-molecule processes on cadmium selenide crystallites. *Russian Journal of Physical Chemistry A*, 1989 vol. 63, no. 6, 839-841.

[78] Smyntyna, V;. Golovanov, V. Analysis of oxygen chemisorption on CdS thin film surface. *Vuoto,* 1994 vol. 23, no1, 86–88.

[79] Smyntyna, V; Kashiulis, S; Mattogno, G; Viticoli, S. Chemisorbed oxygen forms on CdS thin films surface. *Vuoto*, 1994 vol. 23, no.2, 35–37.

[80] Golovanov, V; Smyntyna V; [and others]. Surface Spectroscopy Study of CdSe and CdS Thin-Film Oxygen Sensor. *Sensors and Actuators B*, 1994 vol. 22, 189–194.

[81] Kashiulis, S; Mattogno, G; Smyntyna V; [and others]. Adsorbed Oxygen Forms on CdS Thin Film Surface, Vuoto, 1994 vol. 23, no. 2, 32–34.

[82] Mattogno, G; Viticoli, S; Smyntyna V; [and others] Analysis of Oxygen Chemisorption on CdS Thin Film Surface. *Vuoto*, 1994 vol. 23, no.1, 86–88.

[83] Smyntyna, VA. The chemisorption forms and the center nature of oxygen chemisorption on the CdSe thin film surface. *Il Nuovo Cimento*, 1981 vol. 63B, no.2, 642–650.

[84] Golovanov, V; Smyntyna, V; Mattogno, G; [and others]. Oxygen Interaction of CdS-Based Gas Sensors with Different Stoichiometric Composition. *Sensors and Actuators B*, 1995 vol. 26–28, 108–112.

[85] Smyntyna, VA; Gerasyutenko, VA; Korneva SA. Effect of alloying on the adsorption-desorption sensitivity of films of cadmium selenide. Bulletin of the Academy of Sciences of *the USSR. Inorganic Materials*, 1980 vol. 25, 856-861.

[86] Вашпанов, ЮА; Смынтына, ВА; Сердюк, ВВ. Свойства реальной поверхности тонких пленок селенида кадмия, легированных атомами индия и селена. *Поверхность. Физ. Хим. Мех.,* 1984 вып.1, 93–98.

[87] Smyntyna, V; Gerasutenko, V; Kashiulis, S; Mattogno, G; Viticoli, S. The causes of thickness dependence of CdSe and CdS sensor sensitivity to oxygen. *Sensors & Actuators B, Chemical*, 1994 vol. 19, no.1-3, 464–465.

[88] Смынтына, ВА. Изменение структурных, электрофизических и фотоэлектрических свойств в процессе роста и обработки пленок сульфида и селенида кадмия. *Фотоэлектроника*, 1988 вып. 2б, 58–63.

[89] Бонч-Бруевич, ВЛ. Электронные состояния на неидеальной поверхности. В: *Проблемы кинетики и катализа*, М: Наука; 1970 т.14, 19–27.

[90] Никольский, БП. Редактор. *Справочник химика.* М: Химия; 1964т. 3.

Chapter 5

SEMICONDUCTOR GAS SENSORS

ABSTRACT

Design, composition and characteristics of monoelectrod planar gas sensors are considered in this chapter. Sensors produced by gravure offset printing, electrohydrodynamic spraying as well as by different methods of vacuum deposition, annealing and doping are also investigated. The results of the testing of the various structures as gas sensors are shown. Among the studied materials there are SnO_2, ZnO, PbO, In_2O_3, CdS, CdSe, Si and their complex composites. The surfaces of the simple, binary compounds and complex composites were doped with noble (Au, Ag, Pt) and others (In, Cd, Pb) metals. Macro-, micro- and nano-porous silicones were investigated.

5.1. SEMICONDUCTOR FILM GAS SENSORS

5.1.1. Volumetric One-Electrode Semiconductor Gas Sensors

Volumetric one-electrode semiconductor gas sensor elements applied as gas sensors are ones of the first similar devices. These sensors are obtained in the technological process, which includes high-temperature sintering of semiconductor ceramics by platinum heater unit. These sensors were designed for gas registration and concentration control in the ambient atmosphere and in the technological processes of flammable and harmful gases' presence control, such as hydrocarbons, hydrogen, alcohols, and other similar components [1-8]. Low power consumption and relatively high performance level of volumetric one-electrode semiconductor gas sensors caused their use not only in stationary equipment, but also in portable instruments for detecting and warning leakage of different gases, fire safety systems, etc.

One of possible volumetric one-electrode semiconductor gas sensor design is shown in Figure 5.1 [8], where the platinum coil for heating introduced into the semiconductor ceramic base of gas sensors.

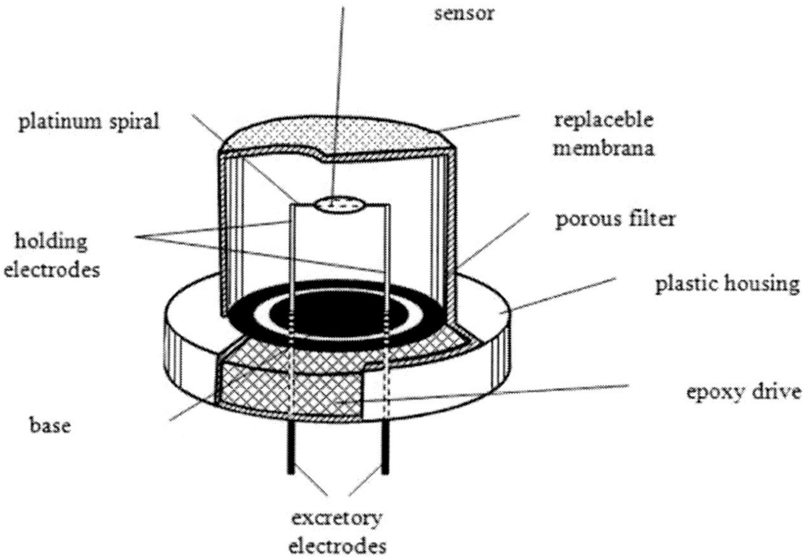

Figure 5.1. Design of a volumetric one-electrode semiconductor gas sensor [8].

The functional concept of one-electrode sensors is based on the shunt of heating resistor by reaction with gas and oxide semiconductor material. Platinum coil acts both as: a heater and a measuring electrode. The sensor operates in constant current mode. When exposed to a gas atmosphere, the sensor output signal is formed as a change in the voltage drop across the sensor.

The change of the resistance at the semiconductor layer leads to the changes in the total sensor resistance *R*, which leads the changes in the released energy RI^2 at constant current mode. One-electrode semiconductor gas sensors are highly sensitive to detectable gases. The adsorption and desorption time of such sensors is a few seconds. The advantage of one-electrode design is the sensor reliability and a small change in their initial settings (≤ 5 %) over the operating cycle.

5.1.2. Planar Semiconductor Gas Sensors

A significant number of planar semiconductor gas sensors are manufactured for a long time, using standard thick film technology, containing gas sensitive material with bonding additives which evaporates during annealing.

Planar semiconductor gas sensor operation is based on the changes of the conductivity of the gas sensitive layer with the composition's changes of the surrounding gas atmosphere [6, 7]. The most commonly used semiconductor materials in such structures are wide-gap semiconductor oxide (SnO_2, WO_3, ZnO), that could be operated at high temperatures (about 770 K).

In a typical design for such semiconductor sensor, the sensitive layer is positioned on a dielectric substrate between two metal electrodes, and a third metal electrode, usually made of platinum, is located on the back side of the substrate and used to heat the element. Such sensor's design requires the independent power supply for the heater and the current transport in the gas-sensitive layer.

One version of the planar one-electrode sensor structure based on SnO_2 is shown at Figure 5.2 [8]. Heater and the oxide film are deposited sequentially on a dielectric pyroceramics substrate. The functional concept of presented sensor is the same as one-electrode sensor, being operated in a mode-current stabilization and is based on the shunt of heating resistor as a result of the interaction of the gas with an oxide semiconductor. When exposed in a gaseous atmosphere, the sensor output signal is formed as a change in the voltage drop across the sensitive element. In the case of reducing gases such as CO, decrease of the total resistance R leads to decreasing power dissipation RI^2 and operating temperature of the sensor.

Despite the fact that the traditional thick-film technology is a fast and economical method for sensors manufacturing, it has some limitations as compared to thin-film technology. For example, the existing technological limitations in achieve a minimum width of conductive lines derived by printed thick-film technology, essentially complicated the application of this technology in the manufacture of integrated sensor structures.

Planar thin-film technology of semiconductor devices manufacturing has the advantage in using it for the line production of sensors, as it allows for one cycle sputtering hundreds or thousands of sensors. Typically semiconductor gas sensors are operating at elevated temperatures. Therefore, each sensor requires its own miniature heater and thermal stabilization system, allowing heat sensitive layer and hold it in the working temperature range.

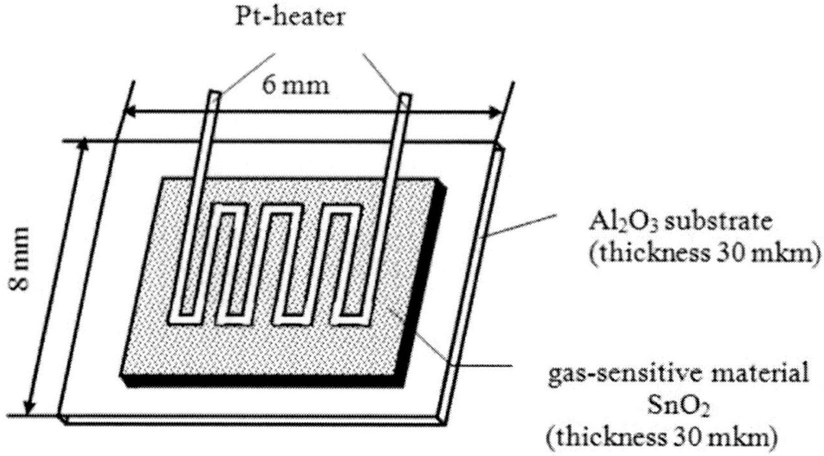

Figure 5.2. Design of a planar semiconductor one-electrode sensor [8].

The *NiCr* and *Pt* films, deposited on the glass-ceramic surface, were annealed at 650 *K*. For creation the mechanically strong contacts, it was used a conductive paste that is based on palladium. In addition, for increasing the strength of contacts placed over the glass-ceramic substrate, the high-power laser beam penetrated the aperture (0.1 *mm* in diameter). Platinum wire was inserted in these apertures and sintered to contact terminal with conductive paste at temperature of 900 *K*.

Obtained sensitive element should be stick to the pad and heated with gold wire.

5.1.3. Gravure-Offset Printing of Semiconductor Gas Sensor

Gravure-offset printing technology as well a thick-film technique that makes it possible to apply the narrower conductive lines as compared to the results obtained by the traditional thick-film technology [9-12]. Using the gravure-offset printing technology, it is possible to reach rather higher packing density of sensor structures. One of the most important advantages of this technique is the possibility of three-dimensional printing.

Form of the required circuit pattern etched on the gravure plate, filled with ink and cleaned up with a special knife so that the ink remains in the etched grooves of the required configuration. Then sample transferred to the substrate by a flexible silicone cushions. Such technology can be applied for conductive lines less than 50 microns in width, which opens the possibility for miniaturization of produced sensory structures. Putting on the small electrodes capability allows to miniaturize the design and to increase the packing density of the sensors. Usage of the gravure-offset printing technology was realized through shorter connection between conductors on a dielectric substrate less than 100 microns in diameter [13-16].

Compared to conventional thick-film technology, the gravure-offset printing technology allows for putting on the planar resistors with much higher resistance per unit area, which allows to obtain thick-film heaters on *Pt* base with resistance about 10^3 *Ohm*.

Usually, heaters are put on the substrate through the use of a commercial platinum solution, modified by addition a suitable mixture of organic binders, in order to obtain correct viscosity of coating material. The typical thickness of the conductive lines produced by this method is depending on the type of solutions and depth of the print mask form.

Lines and interline space width in 50 microns were obtained by putting them on the resistor element on the pyroceramic substrate, this aspect allow to increase the heater resistance by two orders of magnitude and to avoid the *n*-type defects at the initial films doping [8].

5.1.4. Semiconductor Gas Sensors Obtained by the Electro-Hydrodynamic Spraying Method

One of the promising methods for the semiconductor films production is electro-hydrodynamic (EHD) layers deposition in container's acid atmosphere with the use of the various nozzles method which was developed in [17, 19, 21]. The principal feature of this method is that all operations can be carried out in a normal atmosphere. This technology is also characterized by the low equipment's and process operations cost. Layers deposition, such as cadmium sulfide for example, by EHD method, is as follows: process solution prepared by mixing in a certain proportion of alcohol solutions of raw materials, applied in the form of mono-disperse aerosol droplets on a dielectric substrate which is heated up to a high temperature. Aerosol spray mode implemented by applying a high electric potential (about 4 to 6 *kV*) between spray capillary and body of the heating element which contains the substrate. The EHD method allows generating the flow of droplets with micrometer dimensions in high level of mono-dispersion. Charged droplets of process solution drip out from the tip of the capillary and moving to the substrate through the electric field lines forming a divergent cone. By electrostatic repulsion between neighboring droplets it is deposited on the substrate evenly.

Reactive agents are selected by the principle: the reaction products, besides the products required for the formation of the final composition, being volatile at the deposition temperature. Applying the alcoholic solutions [19] of the raw materials instead of the commonly used aqueous solutions of these salts [20], increases the process stability to create a mono-dispersed aerosol what ultimately leads to the forming more developed structure semiconductor film on the substrate.

The promise of this method in producing semiconductor films is defined as a possibility of doping the layer directly in the process of synthesis by introducing the necessary molar concentration of salty ligands in the initial solution. However, the range of doping agent, in this case, is limited by the elements that the salts are not precipitated by pre-mixing with the original solution. This disadvantage can be eliminated by supplying the original and doping solutions in the atomization area through two or more (in the case of greater number of the dopants) capillaries of the isolated vessels. The components, merging in spraying point, provide the chemical reaction between them directly in forming of the semiconductor layer zone.

Figure 5.3 shows an installation diagram for obtaining doped semiconductor films by EHD method. It consists of a mono-dispersing aerosol generator (1), vessels (2) with a process solution, a high voltage source (3), heater (4) which contains the dielectric substrate (5).

The heater power source was used as a direct current device (6), which was equipped with anti-aliasing filter. Mono-dispersing aerosol generator consists of two or more (in the case of greater number of dopants), thin-walled capillaries with an internal diameter in 0.1 *mm*.

The capillaries have obliquely cut ends with the tops that are summarized in a single point which is the source of the electric field lines. This configuration and positions of the capillaries provide the optimal performance mode for mono-dispersing aerosol generator which formed the overall drop of the initial and alloy solutions. Failure to comply with these requirements, results in appearance of the non-overlapping aerosol flames on the account of absence electric field lines of a point source.

Applying the described method for doped films gives several benefits, associated with the process steps number, required for doping, reduce and the possibility of semiconductor layers characterization directly in the process of synthesis, the possibility of doping impurities with which the salts are precipitated by pre-mixing with the original solution and expanding the range of films applying through the creation of spatially separated doped regions in the semiconductor layer [21].

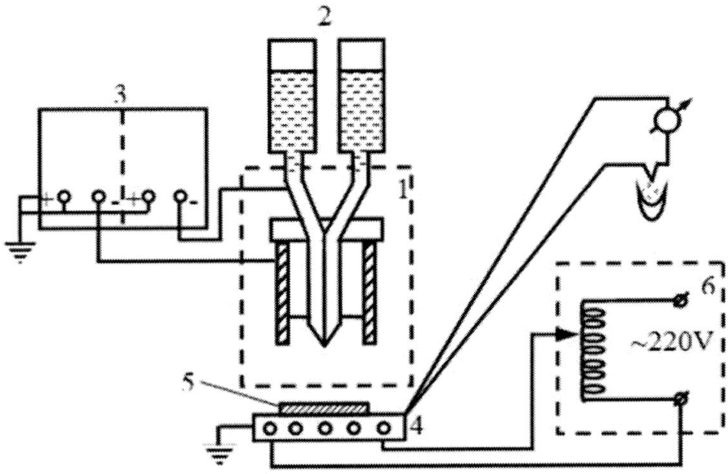

Figure 5.3. Installation diagram for obtaining doped semiconductor films by EHD method [8]. Explanations are given in the text.

Applying the EHD method with the use of the simple technological methods (changing the deposition time, ions ratio in the solution, temperature of the substrate, etc.), the produced semiconductor layers have stoichiometric composition, and the selection character of the dielectric substrate affects the stability of semiconductor films parameters and its dark conductivity, specifically.

Significant advantages of the EHD method of oxide sensors' obtaining, lies in the simplicity of the layer deposits at relatively low temperatures in oxygen atmosphere and ability to apply different activating ligands directly at the obtain layer process. Another advantage of the EHD method is the ability to produce films with structural orientation particles [22, 23], high uniformity and good adhesion quality [22, 24, 25].

Out of the various ways to generate aerosols (pneumatic deposition, ultrasonic scattering and electro-hydrodynamic liquid deposition) in many cases, the EHD method is the preferable one, because it enables to create aerosol droplets with the size of microns' order with a high degree of their mono-dispersion [32]. Due to the small droplets' size and high speed of the chemical reaction, the crystallite sizes of formed films may be very small (less than 500 *nm*) [33].

5.2. Characteristics of Freshly Scattering and Annealed Cadmium Selenide Films as a Sensor for Oxygen Concentration Testing

Directly, after the *CdSe* sensors thin-film deposition on the substrate, the stabilization of the residual pressure in process volume and temperature of the produced sensor at $T = T_r$, (room temperature) the pressure increase of dry oxygen is accompanied with increasing the resistance of the sensor, starting up at $p > 10^{-4}$ *mm Hg. Art.* (Figure 5.4) and is particularly essential at $p > 10^{-1}$ *mm Hg. Art.* [26].

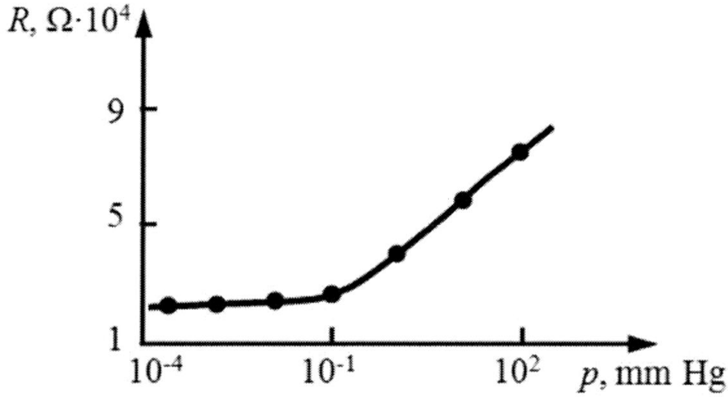

Figure 5.4. Dependence of the freshly prepared *CdSe* layer resistance from the pressure of dry oxygen. Measuring temperature is a room temperature.

Due to interaction of oxygen, existing in air, with the sensor at atmospheric pressure and room temperature the sensors resistance increases by two orders of magnitude and is stabilized after 100-120 minutes after the depressurization of process chamber. Increasing the operate temperature of the sensor leads to exponential increasing in the rate of chemisorption resistance growth of the sensor (Figure 5.5, curves 1-4). It provides the sensors with the identical initial parameters, obtained in a single process cycle.

Simultaneously, the increase in the rate of resistance growth as a result of oxygen chemisorption at T increased up to 150 °*C*, also the adsorption increase of the sensor sensitivity occurs, which is calculated by the simplified formula $\beta = (I_r - I_k) / I_k$ (Figure 5.6), here I_r, I_k - initial and final current value.

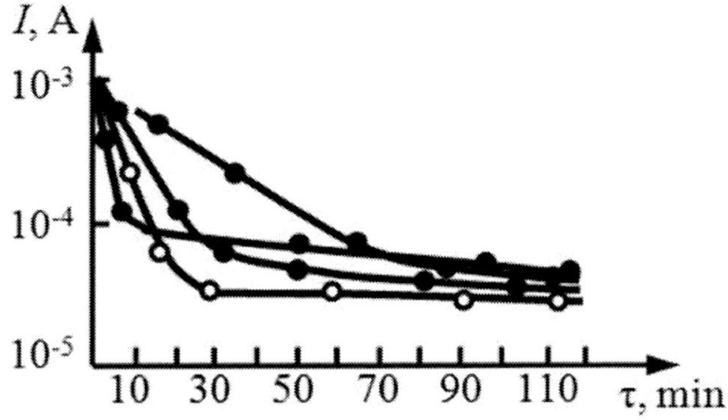

Figure 5.5. The chemisorptional decrease of the current, which flows through the freshly prepared *CdSe* sensor. Temperature measurements are: 1 – 100, 2 – 110, 3 – 135, 4 - 150 °*C*. The sensors are provided with identical initial parameters obtained in a single process cycle.

Increasing in *T* value above 150 °*C* leads to a further increase in the rate of oxygen chemisorption [33], but the β is decreased at same time (Figure 5.6).

Studies of the reproducible sensor parameters at multiple changes of the pressure value in atmosphere which contains oxygen found that at $T = 150$ °*C* after some decrease the amount of β its value stabilizes after 5 process cycles (Figure 5.7, curve 1). But at 200 °*C* (curve 2) β with increasing of the process cycles number its value decreases monotonically.

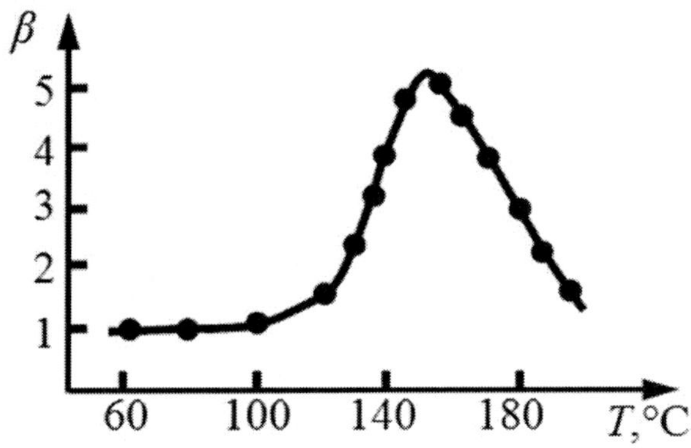

Figure 5.6. Dependence of the adsorption sensitivity $\beta = (I_r - I_k) / I_k$ of a freshly prepared *CdSe* sensor on temperature values.

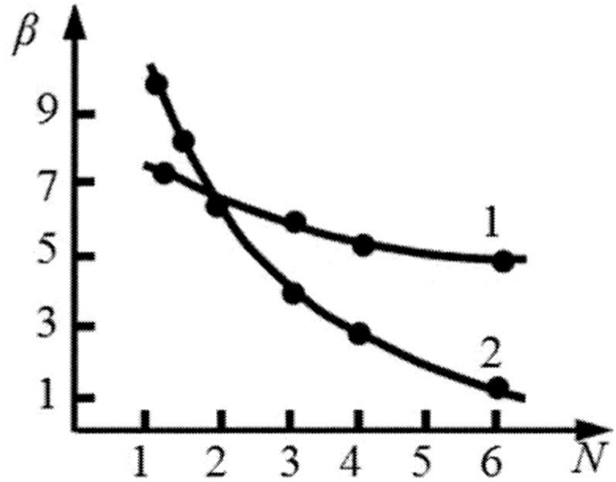

Figure 5.7. Dependence of the adsorption sensitivity β of freshly oxygen sensor on the process cycles number at 150 °*C* (1) and 200 °*C* (2).

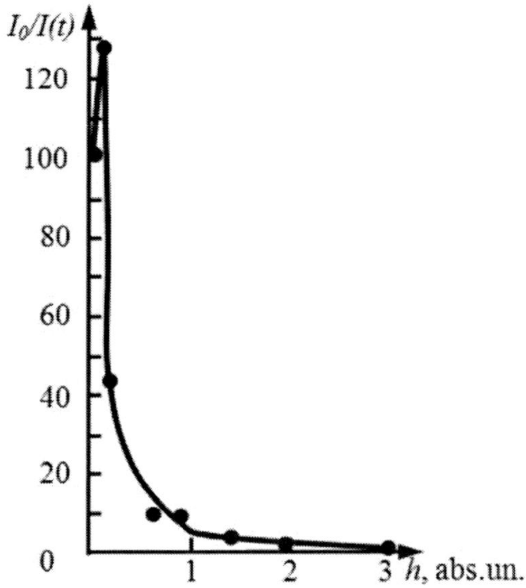

Figure 5.8. Dependence of the adsorption sensitivity of a freshly prepared *CdSe* oxygen sensor on the sensors' thickness.

This effect can be considered as a degradation which is caused by at least two basic physical processes [27]:

a) the diffusion of oxygen atoms in the sensor volume for interlayers in crystal;

b) the recrystallization, which contained a considerable number of prepared sensors and modifying the structure of the cubic to hexagonal.

The greatest adsorption sensitivity β is presented in the thinnest sensors (Figure 5.8). As the sensors thickness increase the β value decreases monotonically and the thickest of these sensors actually insensitive to oxygen.

The thinnest *CdSe* film sensors are finely dispersed (Figure 5.9, a) and consists of crystallites with a mixed structure: cubic and hexagonal [27]. By increasing the layer thickness, the crystallite size is increasing (Figure 5.9, b). At the same time there is a change in their structure: it is a monophasic - hexagonal in result of recrystallization [27].

Thus, reducing the β value with increasing sensor thickness, except the geometric expansion of conducting channel [28] to be able to determinate by a number of physical reasons that caused by peculiarities of sensor film growth:

- reduction of surface interaction with oxygen and *CdSe* sensors by increasing the size of the crystals;
- change of the crystal structure type;
- change of the surface composition of the crystallites due to the recrystallization during the growth.

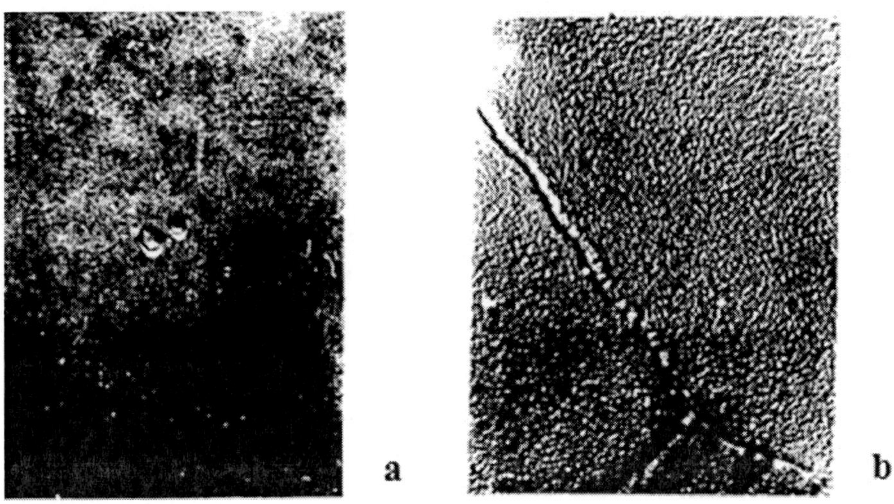

Figure 5.9. Electron-microscopic picture of the surface of thin (a) $h = 0{,}1$ *mkm* and thick (b) $h = 3$ *mkm* freshly prepared oxygen *CdSe* sensors (x 50000).

Freshly prepared *CdSe* sensors demonstrate high adsorption sensitivity to oxygen (Figure 5.6), but its magnitude is relatively rapidly decrease with increasing of process cycles number (Figure 5.7) due to other reasons than above indicated and the dissociation of O_2 molecules into oxygen atoms which diffuse into the sensors volume and leads to an irreversible increasing in its resistance [27].

Heat treatment in air at temperatures higher than operational one leads to complete recrystallization of the sensor structure and saturation of diffused oxygen, thus eliminates the fast degradation of β (Figure 5.10).

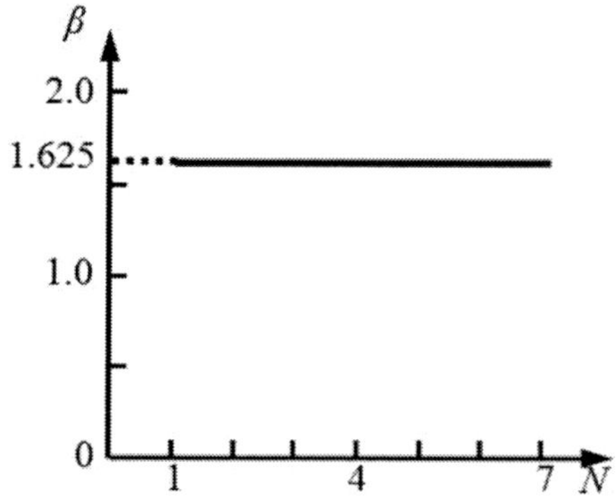

Figure 5.10. Dependence of the adsorption sensitivity β of the recrystallized *CdSe* sensor on the number of process cycles. Measuring temperature is 200 °*C*.

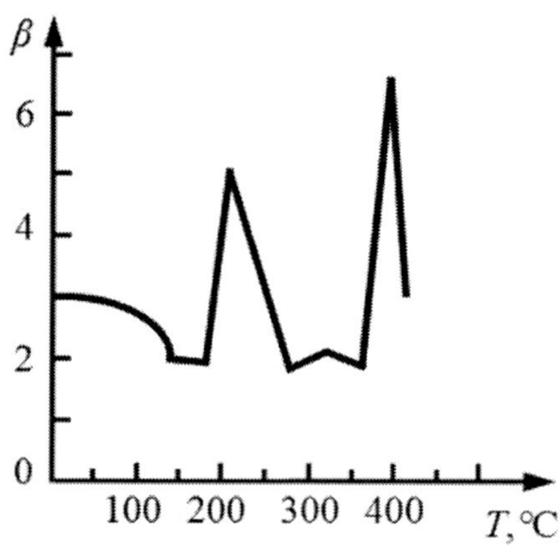

Figure 5.11. Changes of the *CdSe* sensors adsorption sensitivity β depending on the temperature of its treatment. Measuring temperature is 200 ° *C*.

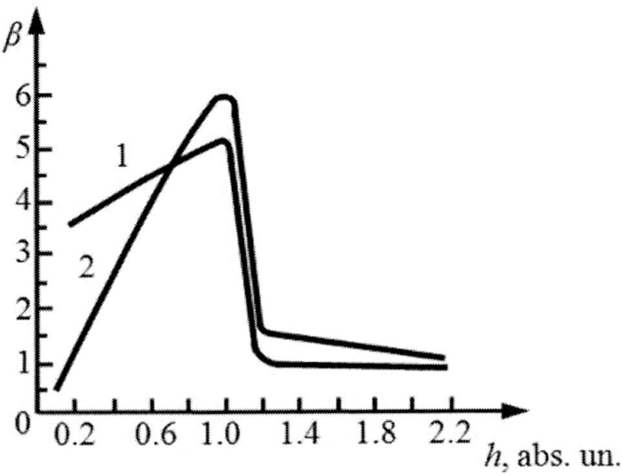

Figure 5.12. Changes of β depending on thickness increasing of *CdSe* sensors that annealed at 450 ° *C* (1) and 500 ° *C* (2).

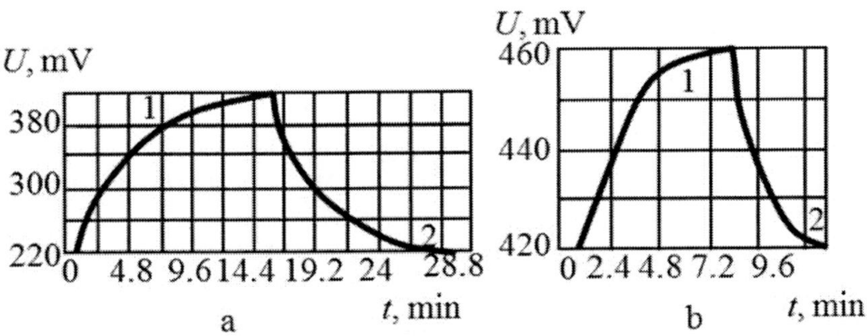

Figure 5.13. Dynamics of the signal changes at the load resistance in the measuring circuit of *CdSe* sensor when concentration of O_2 changes from 15% (1) to 23% (2). The measurements were performed with T_1 (a) and T_2 (b) (T_2 - T_1 = 10 °*C*)

Adsorption sensitivity β varies not monotonically with increasing of the annealing temperature (Figure 5.11). Discovered two temperature ranges after annealing at which β reaches maximum values.

As a result of high-temperature annealing it was found that the highest β value possesses not the thinnest *CdSe* layers and the sensors with an intermediate thickness (Figure 5.12). It is caused by changes in the structure and composition of the surface and near-surface region of sensors, and the oxidation-reduction processes which analysis and mechanisms interpretation are described in [27, 31].

As the sensors operating temperature is increased, the time value of oxygen adsorption τ_a and desorption τ_d are aligned (Figure 5.13). If, at the relatively low operating temperatures τ_d is much more then τ_a (Figure 5.13, a) so in the region of more high *T*, sensor achieves the conditions under which τ_d is near in magnitude to τ_a (Figure 5.13 b).

5.3. Oxygen Sensors Based on *CdSe* Films with the Indium-Doped Surface (*CdSe : In*)

Analysis of the predicted and experimental research results of the biographical semiconductor characteristics influences on the surface properties of a sensor enable to mark out some of the basic factors that are responsible for the adsorption properties: concentration and biographical character of surface defects; concentration and character of dopant atoms placement on the surface. It makes possible to determine the technological parameters that provide the most significant influence on the sensor properties forming in the process of growth and doping. Considering the results of [27] together with the previous sections of this chapter it possible to conclude that the main technological parameters are: the deposition time of semiconductor sensor (thickness *d*), substrate temperature, the temperature and duration of annealing process before doping, the spraying time of the dopant (dopant concentration on the surface) and the type of dopant.

The main performance properties of the sensor that must be controlled in the process are: the adsorption sensitivity β, the adsorption time τ_a, the desorption time τ_d, stability ($\Delta = \Delta U_d / \Delta U_i$; ΔU_i - signal changes by the oxygen concentration changing, $\Delta U_d = \Delta U_i - \Delta U_{i+1}$).

Studies of the dopant character influence on the sensors characteristics [27] have shown the advantage of indium in comparison with other materials. During the researches [29] were determined the optimal thickness and annealing modes for based on cadmium selenide films sensors [30].

Statistical treatment of the results of the process parameters influence on the sensor characteristics and compared it with each other showed that for not annealed cadmium selenide films optimal process mode is $T = 570\ K$ and doping time is 30 *s* (Table 5.1, Table 5.2.).

But in some parameters such samples able to yield to films obtained in other process modes. It allows amplifying the certain positive qualities of the sensor in the dependence for needs by varying process conditions.

Table 5.1. Results of research data statistical treatment of not annealed *CdSe: In* sensors

T, K \ t, s	10	20	30	40	60
420	26,8	10,7	11,1	23,1	26,7
470	17,8	16,5	17	15,2	22,2
520	-	-	6	22,1	26,8
570	39,2	24,3	22	21,4	20,2

Table 5.2. Results of research data statistical treatment of annealed *CdSe: In* sensors

T_p, K \ t, s	10	20	30	40	60
420	40	31,2	-	-	-
470	15,2	11	4,0	9,5	13,3
520	11,3	19,8	12,7	10,6	11,2
570	19,1	28,6	10,3	12,4	27,2

For example, in the case of annealed films maximum sensitive $\beta = 0{,}1$ % reached at $T = 470$ *K* and $t = 10$ *s*, the minimum adsorption time is obtained if $T = 520$ *K* and $t = 40$ *s*, better stability belongs to the sensors that are obtained at $T = 570$ *K* and $t = 30$ *s*.The sensors calibration characteristic (NSIQ - normal static instrument quality in accordance with GOST 20220) obtained by consecutive five times passing of five gas mixtures (N_2, O_2) that are characterized by the concentration of O_2 in the range (15-30) % (Tab.5.3).

It should be noted the linearity of the calibration sensor characteristics which is an essential positive quality of the element that simplifying the other blocks structure of gas analyzer.

Dynamic characteristics (τ_a, τ_d - adsorption and desorption time respectively) were determined in accordance with GOST 20220 as setting time of 0.95 stationary signal value.

Measurements are shown that advanced technology allows to obtain about 50 % of sensors with $\tau_a = (2\text{-}4)$ and $\tau_d = (3\text{-}5)$ minutes from one simples group that makes it suitable for practical application in gas analyzers. At the optimal operating temperature of the sensor (350-355) *K*, the duration of its work in unset-up mode is (50-100) hours (the time during which the inaccuracy in the O_2 concentration measurement does not exceed 10 %).

Table 5.3. The sensor signal value which taken from the load resistor in series-connected to the sensor at different oxygen concentrations

Measuring number	Oxygen concentration %				
	15	18	21	27	30
	Output signal value U, 10^{-3} V				
1	3,62	3,41	3,18	3,06	2,82
2	3,54	3,52	3,24	3,07	2,75
3	3,73	3,46	3,25	3,02	2,6
4	3,63	3,48	3,2	3,1	2,82
5	3,66	3,45	3,25	3,1	2,79
Average value	3,64	3,46	3,22	3,08	2,76

Basic reduced error in the oxygen concentrations analysis of N_2 and O_2 mixtures in sensor, as calculated in accordance with GOST II. 002-73 is δ^0 = 3-8 % of capped rate in measurements.

Complementary error of analysis in O_2 concentration by obtained sensors derived the influence of unmeasured components CO_2 and H_2 are 1.14 % and 2.84 %, respectively [27].

Table 5.4 contains the basic parameters of the known oxygen analyzers. In the last column shows the sensor parameters based on *CdSe:In* films [27].

In comparison with the shown on Table 5.4 the *CdSe:In* sensors distinguish favorably to the power consumption, technological dimensions, weight and the prospects of miniaturization. In addition, the cost of *CdSe:In* sensors is much lower than the cost of serial industrial devices.

In some industries, where the precision provided by *CdSe:In* sensors is sufficient, they are able to apply a cheap and compact instrumentation that developed on the basis of semiconductor sensors are protected by copyright certificates on inventions [31, 32, 33].

Table 5.4. Comparison table of oxygen analyzers

N	Method, device, parameters	Chemical method	Solid electrolytes	Thermomagnetic method	Thermo-induction metrical method	Mass-spectrometry method	Luminescent-kinetic method PL 5101	*CdS :In* sensors
1	Range	0–10 %	0,25 %	0–100 %	0–30 % sensitivity	0–25 %	$4*10^{-8}$–10^{-5} %	0,01–30 %
2	Error	0,3 %	1–5%	1%	0,1%	2%	0,10%	6–8 % of capped rate in measurements
3	Components analysis time (persistence)	15–20 minutes	10–20 *s*	10 *s*	15–40 *s*	10 *s*	2,5 minutes	τ_a = 2–3 minutes, τ_d= 3–5 minutes
4	Selectivity	selective	about 1 % H_2	selective	about 2 % H_2	selective	selective	about 0,5 % H_2
5	Time of continuous operate (without tuning)	30 *h*	500 *h*	–	1–15 days			100 *h*
6	Size	216x520x620 mm^3		560x278x229 mm^3			400 *kg*	sensor 5x5x0,2 mm^3 sensor 0,5 g
7	Consumed. power					14 *kW*		sensor 10*W*
8	Annotation	Device of discrete operation	Temperature 1000 *K*			Device of discrete operation		

5.4. Oxygen Sensors Based on CdS Films and Metal Oxides

5.4.1. CdS Oxygen Sensors

Applying of electro-hydrodynamic liquid spraying enables to produce *CdS* sensors for register the oxygen concentration in surrounding atmosphere [8].

The optimum temperature that able to obtain *CdS* sensors with the highest β value is in the (670-700) *K* range (Figure 5.14).

At lower T_p low values of β are caused by presence of un-sublimated NH_4Cl in the crystals inter-layers [39]. Hygroscopic ammonium chloride absorbs the moisture from the surrounding atmosphere and shunts by ionic conductivity the high-resistance area that created by chemisorbed oxygen at the sensor surface.

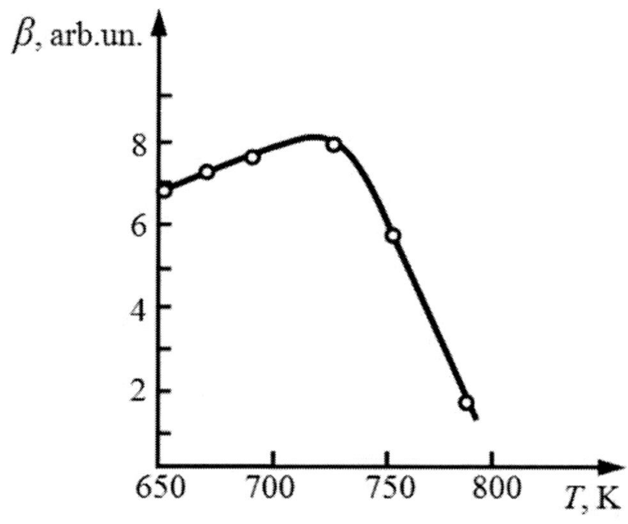

Figure 5.14. Dependence of the *CdS* sensor sensitivity to the O_2 adsorption on the substrate temperature during the deposition of semiconductor layer. Operate temperature of *CdS* sensor is 420 *K* at oxygen chemisorption.

Increasing in T_p value above 700 *K* stimulates forming of the *CdO* conductive layers [41], and it also brings to decreasing in value of β.

Field of the minimum times values of adsorption τ_a and desorption τ_d the oxygen to *CdS* sensors is in (330-340) *K* temperature range (Figure 5.15).

Practically linear dependence of the adsorption sensitivity β of CdS sensors on the oxygen concentration in (5-40) % range (Figure 5.16) enables to use it as sensitive elements of gas analyzers.

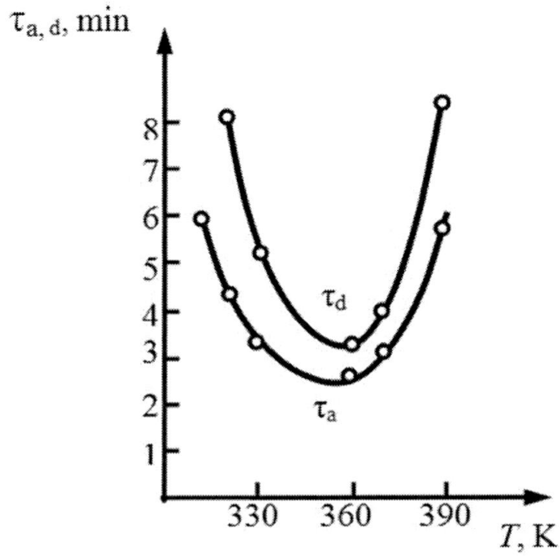

Figure 5.15. Temperature dependence of the adsorption and desorption time of *CdS* sensor in oxygen chemisorption presence.

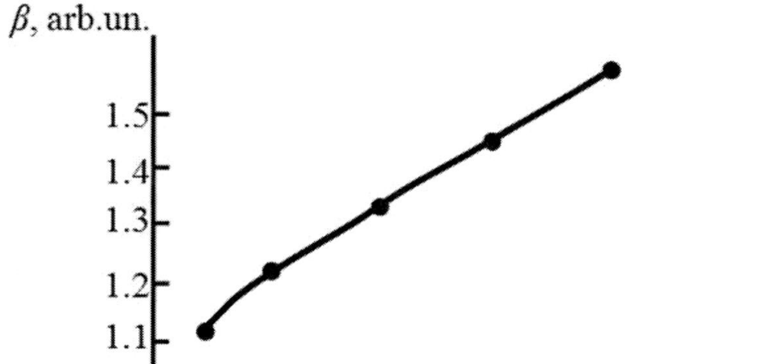

Figure 5.16. Dependence of the adsorption *CdS* sensor sensitivity on the oxygen concentration. Operate temperature is 360 *K*.

Increasing of the *CdS* oxygen sensors adsorption sensitivity β about 5 times is achieved by aluminum doping directly in the manufacturing process with spraying the alcoholic solution of aluminum chloride through the additional capillary [8]. This eliminates the process steps required for doping the sensor after its production.

5.4.2. Oxygen Sensors Based on Doped Metal Oxides

Oxygen sensors based on metal oxide films usually obtains by high-frequency sputtering of accumulation electrode in the thin oxygen atmosphere [36], in which the oxidation degree of the metal adjustable by oxygen pressure in the process chamber. With this method by a single cycle it is possible to obtain the mixed metal oxides and dope it with the purpose to increase the adsorption sensitivity β.

Technological researches showed that the most suitable for practical application are the oxygen sensors based on zinc oxide that included a mixture of $ZnO{:}SnO_2$ and $ZnO{:}In_2O_3$ oxides with doped by indium surfaces: $ZnO{:}In_2O_3 / In$, $ZnO{:}Sn_2O_3 / In$ (Table 5.5) [36].

The $ZnO{:}In_2O_3$ / In (1:3) oxygen sensors showed the smallest measurement error (about 0.6 %) and the highest service life (about 1100 hours), and the most sensitive (about $6 \cdot 10^{-1}$) are the $ZnO{:}Sn_2O_3 / In$ (1:2) sensors.

Nominal static characteristics of the metal-oxide oxygen sensor that operates at 353 K are presented in Table 5.6.

In these conditions sensors adsorption time was τ_a= 48 *s* and desorption time was τ_d = 102 *s*.

The value of the basic reduced error (GOST 11.002-73) δ_{os} depends on the oxygen concentration (Table 5.7), and its value is able to change by two times when the oxygen concentration changes at the middle of the interval (δ_{os} – the minimum) to the borders of the measurement interval (δ_{os} – the maximum).

Changes of the oxygen partial pressure value in the measuring volume of 65 *kPa* to 287 *kPa* leads to disastrous errors in measurement at the lowest oxygen concentration which used. At the same time with used the maximum oxygen concentration the same changes in the oxygen partial pressure value leads to an error in the measurements of the sensor signal about not exceed 10%.

Table 5.5. Dependence of adsorption sensitivity and service life of the oxygen sensor on the metal components ratio in the composition obtained by high-frequency sputtering

Sensor composition	Adsorption sensitivity	Error %	Service life, *h*
$ZnO:SnO_2$ / In (1:1)	2 X 10^{-2}	2	700
$ZnO:In_2O_3$ / In (1:1)	4 X10^{-1}	1,9	700
$ZnO:In_2O_3$ / In (1:2)	6 X10^{-1}	1,8	800
$ZnO:In_2O_3$ / In (1:3)	2 X 10^{-1}	0,6	1100
$ZnO:In_2O_3$ / In (1:4)	5 X 10^{-1}	1,2	700
$ZnO:In_2O_3$ / In (1:5)	2 x 10^{-2}	4	600

Table 5.6. Nominal static characteristic of the metal-oxide oxygen sensor (*$ZnO:In_2O_3/In$*)

	Output signal value *U, mV*							
N/c, %	0	5,1	7,8	10,2	17,3	21,8	27	31,2
1	0	0,21	0,66	1,11	2,43	3,42	4,23	4,31
2	0	0,22	0,62	1,08	2,38	3,34	4,24	4,33
3	0	0,23	0,64	1,1	2,4	3,44	4,26	4,35
4	0	0,23	0,67	1,08	2,33	3,31	4,11	4,3
5	0,01	0,22	0,64	0,11	2,39	3,37	4,15	4,31
At an average	0	0,22	0,646	1,098	2,386	3,376	4,198	4,32

Additional error in measuring of oxygen-containing atmosphere appears due to the presence the maximum allowable concentration of a gas from a number of CO, CO_2, NO, NO_2, H_2, H_2O, NH_3, CH_4, *Freon-22*, acetone, it significantly effects on the ZnO: In_2O_3 / In sensor reading only at minimum of oxygen concentration range of the all used range.

Among acceptor gases the greatest measurement error of changes in the oxygen concentration caused by presence the maximum allowable concentrations of NO (δ_{NO}= 0,46 %) and NO_2 (δ_{NO2} = 0,7 %) in measured atmosphere and among donor gases it is the water vapor (δ_{N2O} = 0.54 %) [36].

Table 5.7. Dependence of δ_{OS} value of the $ZnO:In_2O_3/In$ sensor on oxygen concentration

Oxygen concentration%	5,1	7,8	10,2	17,3	21,8	27	31,3
δ_{OS} %	0,91	0,49	0,43	0,41	0,41	0,71	0,9

Almost the linear change of β on the O_2 concentration in the (0-30) % range (Figure 5.16) has a cadmium sulfide layer as a low-temperature oxygen sensor.

5.5. *CdS* Sensors for SO_2 Detection

Sulfur dioxide is the one of most common air pollutants. Traditional gas analyzers for sulfur dioxide are being based on different principles: thermal, electrochemical, fluorescence, optical, photo- calorimetric, spectrum-absorption and others [36, 37]. Selection of a certain gas detection method depends on the operating conditions and problems set. Among the relatively new gas analysis methods based on using the solid-state sensors, promising is an application of gas molecules adsorption on the semiconductor films surface which leads to change in the conductivity and / or other layer parameters [38-66].

Development of these semiconductor sensors based on thin semiconductor films allowed applying the microelectronic planar technology for its production and it creates on a single substrate the measuring systems: sensor - microprocessor, i.e. to produce smart sensors. Development of this area is one of the most urgent problems of modern instrument-making.

General requirements for semiconductor sensors that are used as operating sensors are following: semiconductors must be sufficiently chemical-resistant in the operating temperatures, pressure and concentration fields, not to form stable compounds with adsorbed particles and possess sufficient mechanical and thermal strength. The most widely distributed sensors that based on metal-oxide semiconductors such as SnO_2, TiO_2, ZnO, Fe_2O_5, V_2O_5, CdO, and others [38-43, 67]. However, its possess several drawbacks, in particular - high interference to prevented components in the testing environment, and significant adsorption and desorption times. Performance increase of sensors that based on these structures provides by increasing of the operating temperature. Thus, oxygen detector that based on nickel oxide has acceptable values of adsorption and desorption times at 1300-1500 *K*, and sensor that based on *CO* at 1100-1300 *K* [66-69]. At these temperature ranges along with the electron-molecular phenomena intensive developed the degradation processes that stimulated by high temperature. These processes cause the deviations in sensors characteristics and its outlet from a range of operating values. Also, there is a low degree of composition and properties reproducibility of the semiconductor oxide-metal films surface even at identical operating conditions. So, quite competitive are the thin-film semiconductor sensors that based on low-temperature electron-molecular processes on the surface, in particular - cadmium sulfide film sensors obtained by electro-hydrodynamic spraying method. A well-developed surface of *CdS* sensors makes it very sensitive to the composition changes of surrounding gas environment. In the interaction of indicated *CdS* sensors with SO_2 its resistance changes in 2-10 times. [8]

When sulfur dioxide implements chemisorption on semiconductor surfaces, like O_2, it exhibits acceptor properties [69] and leads to a decrease of the sample electrical conductivity, and due to the active interaction with the semiconductor material SO_2 occupied a higher position than oxygen that caused by the greater value of the oxygen electron affinity than with O_2 [70, 71]. Therefore, the chemisorption of sulfur dioxide in the studying cadmium sulfide thin films may provide some effects on the current transport in these layers; it is similar to the effect of chemisorbed oxygen.

Since sulfur dioxide is high heat-resistant and dissociation of its molecules is not observed even at 1770 *K* [71] than with applying the temperature conditions of 350-450 *K* SO_2 chemisorbed in molecular form. Significant size of sulfur dioxide molecules prevents diffusion in volume of *CdS* sensor layer. Therefore chemisorption of SO_2 molecules realize primarily on the geometric surface of sensors. This is indirectly confirmed by the fact that the *CdS* sensor adsorbed sensitivity to SO_2 dependence on the thickness (Figure 5.17, curve 1) is similar to the same dependence of β on O_2 chemisorption, which realized on geometric surface at $T \approx 330$ *K* (Figure 5.17 , curve 2).

Under natural conditions the sensor for SO_2 registration operates in atmosphere which contains oxygen. In addition, for the practical use sensor must be able to detect changes of the concentration value of sulfur dioxide gas in the field of (1-15) *ppm*, which is about a hundred thousand times smaller than concentration of free oxygen in the atmosphere. Since the oxygen and sulfur dioxide adsorption influenced on the sample conductivity the compete acceptor action, than selection of the optimum operating temperature of semiconductor sensor and conditions of its synthesis must provide the necessary sensitivity and selectivity of the *CdS* sensor to SO_2 along with minimal time of transient analysis.

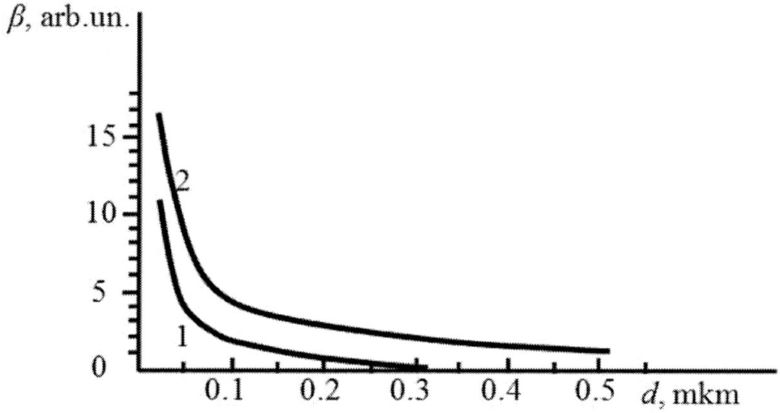

Figure 5.17. Dependence of the *CdS* sensor sensitivity on its thickness in case of sulfur dioxide chemisorption at 360 *K* (1) and oxygen chemisorption at 330 *K* (2).

Figure 5.18 shows dependences of *CdS* sensors adsorption sensitivity β to the 20% oxygen concentration (curve 1) and 9 *ppm* sulfur dioxide concentration (curve 2) on the operating temperature. As follows from this figure, in the whole studied temperature range the *CdS* sensor has significant sensitivity to oxygen with a maximum at 440 *K*. The highest adsorption sensitivity to SO_2 chemisorption is observed in the range of 350-360 *K* where value of its β to O_2 is negligible. At $T > 570$ *K* adsorption sensitivity β of sensor become less than one, and SO_2 provides donor action, leading to an increase in the sample conductivity.

This fact made by displacement of the Fermi level in the band-gap with temperature increasing below the level of the chemisorbed sulfur dioxide molecules energy (ΔE_a) in the band-gap, which leads to a change in the equilibrium state between different forms of chemisorption [75].

Applying the temperature (~ 570 *K*) and the dark current value, which provides transition from the acceptor effect of SO_2 chemisorption to the donor effect, as shown in [8], made an assessment in band-gap the *CdS* energy level formed by molecule of sulfur dioxide, its equals 0,35 ± 0.05 *eV* of the conduction band bottom.

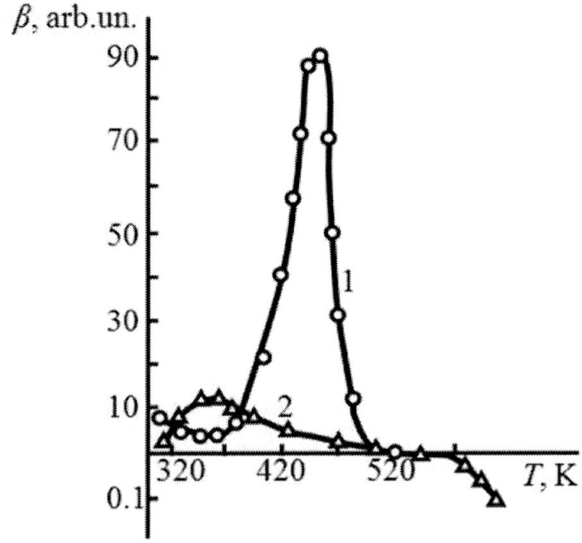

Figure 5.18. Dependences of the *CdS* sensors adsorption sensitivity β to 20% oxygen concentration (curve 1) and 9 *ppm* sulfur dioxide concentration (curve 2) on the operating temperature.

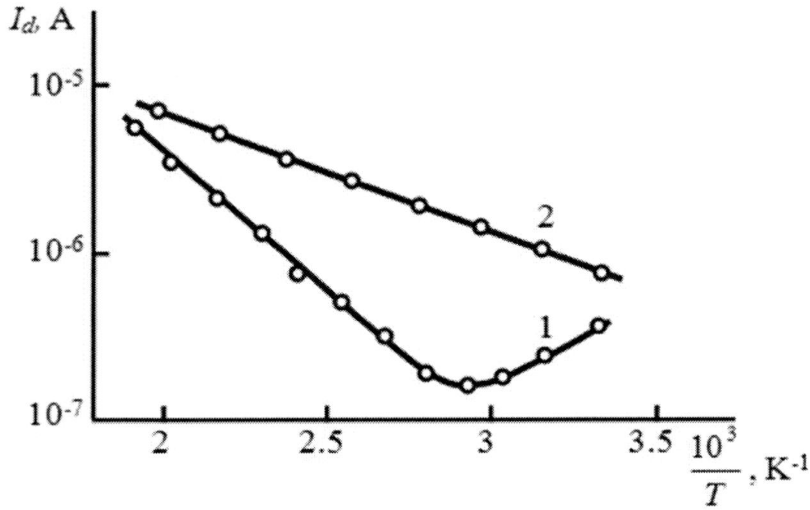

Figure 5.19. Temperature dependence of *CdS* samples dark current that measured in an inert nitrogen atmosphere (2) and a mixture of nitrogen and sulfur dioxide in concentration of 9 *ppm* (1).

As follows from the Figure 5.19, the character of SO_2 chemisorption is activation, as evidenced by a decrease of conductivity with temperature increase in the initial stages of heating (Figure 5.19, curve 1).

The growing plot of the curve 1 (Figure 5.19), caused by the desorption of sulfur dioxide and the activation energy of conduction defined in this plot related to the electrons emission in the conduction band from the chemisorbed SO_2 energy levels, and value of the corresponding energy depth level (ΔE_a) is 0.38 eV. The resulting value is in good agreement with the value of $\Delta E_a = 0.35$ eV that found from measuring the temperature dependence of the films sensitivity to sulfur dioxide (Figure 5.17) and described in [8] by assessment calculations.

Presence of the SO_2 donor effect on *CdS* sensors at $T > 570$ K (Figure 5.18, curve 1), suggesting that for elimination of the oxygen effect on the measurement of SO_2 in the oxygen-containing atmosphere should increase the operating temperature of the sensor above 570 K. However, at these temperatures, there is a significant degradation of characteristics and irreproducibility of sensor parameters at repeated SO_2 influence. Therefore, from a practical option the optimum operating temperature is (360-390) K, at which the degradation of sensor parameters is insignificant, and the adsorption and desorption times are small (10-150 seconds).

The maximum value of β achieved if the substrate temperature is in the field of 670 K at the putting sensor to it (Figure 5.20). Sensitivity decreasing at a lower substrate temperature, as in case of *CdS* sensors for oxygen (section 5.4.1), may be related to the presence in granular interlayers the unsublimated ammonium chloride (NH_4Cl) [27]. Decrease of β with increasing temperature above 700 K caused by *CdS* sensor shunt that formed by conducting cadmium oxide layers [29].

Dependences of adsorption τ_a and desorption τ_d times on the temperature (Figure 5.20) obtained by measurements of the relative values of the sensors dark current (I/I_0) with time at different temperatures after SO_2 inlet to the measuring chamber (Figure 5.22). The advantage of *CdS* sensors in compared to metal-oxide semiconductor layers, that most widespread now as sensitive elements of semiconductor gas detectors [40-43, 74], is the high occurrence rate of the adsorption-desorption processes. As Figures 5.21 and 5.22 shows, by selection of the sensor operating temperature possible to reduce the adsorption time to 10 seconds, and desorption time - up to (40-60) seconds. Small values of τ_a and τ_d also evidence the fact that the molecule of sulfur dioxide chemisorbed on the geometric surface of the films and does not penetrate into the layer depth.

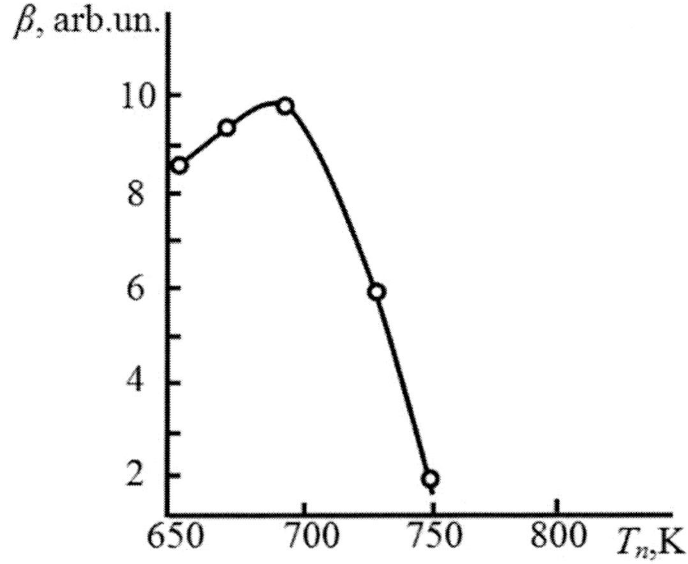

Figure 5.20. Dependence of the *CdS* sensor adsorption sensitivity β to the SO_2 adsorption on the substrate temperature during the deposition of semiconductor layer. Temperature of *CdS* sensor at SO_2 chemisorption is 360 *K*.

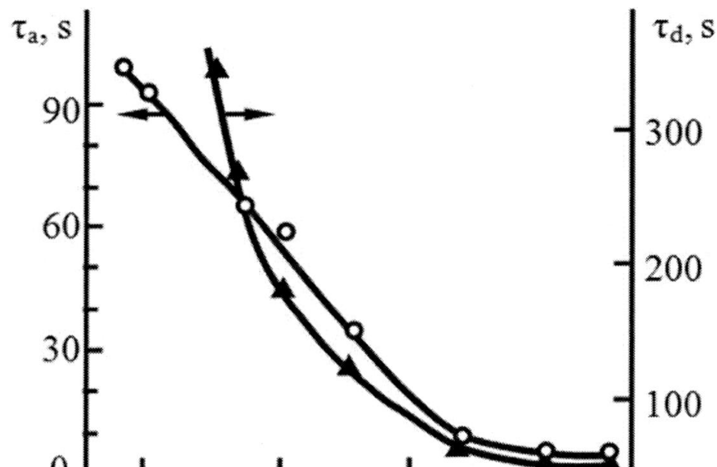

Figure 5.21. Dependence of the adsorption τ_a and desorption τ_d times of SO_2 molecules from the *CdS* sensor surface on the temperature.

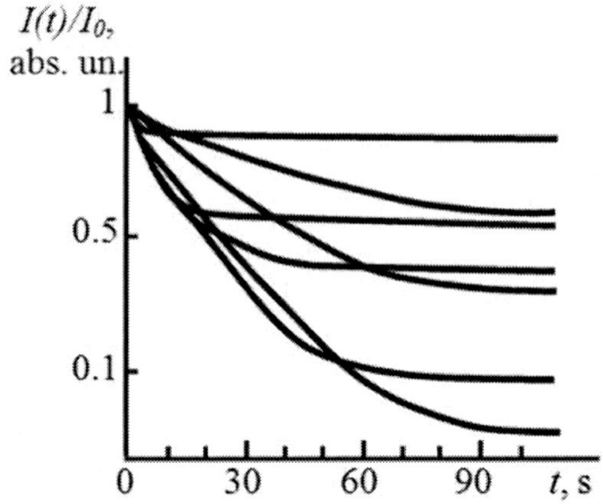

Figure 5.22. Changes in the relative values I/I_0 of *CdS* sensors dark current on time after the SO_2 inlet to the measuring chamber at the different temperatures: 310 *K* (1), 320 *K* (2), 360 *K* (3), 370 *K* (4), 400 (5), 440 *K* (6), 470 *K* (7).

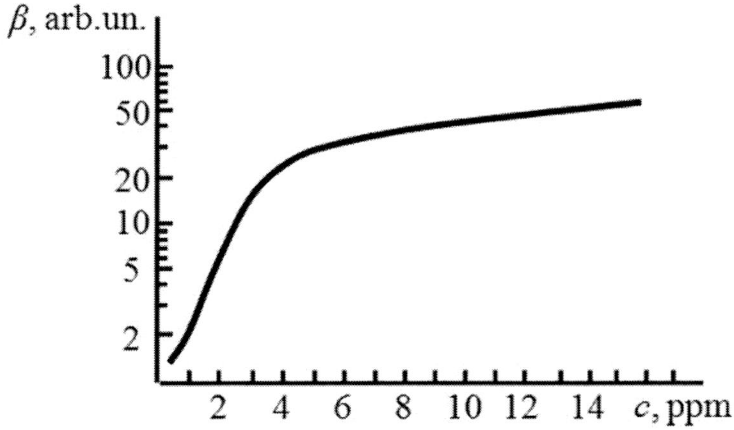

Figure 5.23. Dependence of the *CdS* sensor adsorption sensitivity β on the SO_2 concentration. Operating temperature is 370 *K*.

In the SO_2 concentrations range of (0-4) *ppm* the adsorption sensitivity β of *CdS* sensors to chemisorption of sulfur dioxide is a linear function of this gas concentration (Figure 5.23). In this case the measurement error was 6% of the full concentrations range. As far as fill the surface of *CdS* sensor by chemisorbed SO_2 molecules at concentration greater than 4 *ppm* the β dependence on the sulfur dioxide concentration deviates from linear.

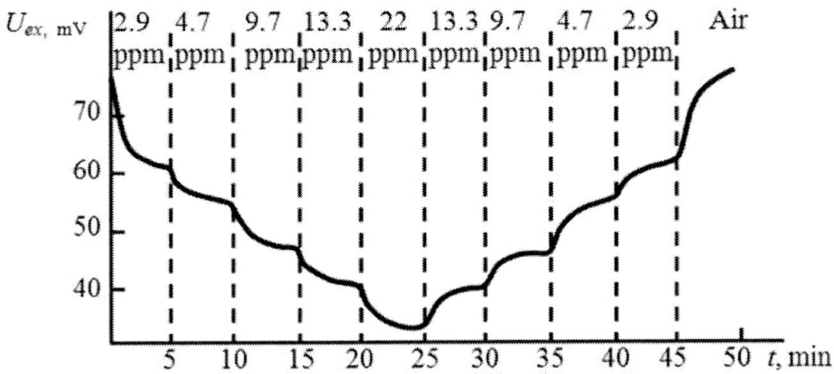

Figure 5.24. Dependence of the output signal of the *CdS* sensor on time with sequential increase and decrease of SO_2 concentration in the air.

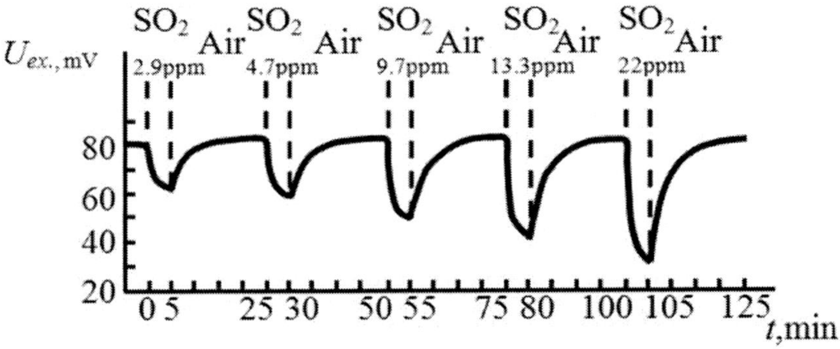

Figure 5.25. Dependence of the *CdS* sensor signal on time with sequential increase of the sulfur dioxide concentration and consistently blown by clean air.

During the operation mode the adsorption and desorption times of *CdS* sensor at T = 370 K increased to (90-300) s with increasing of the SO_2 concentration (Figure 5.24). After two or three effects by maximum SO_2 concentration on the *CdS* sensor is achieved within the measurement error the reproducible of source output signal in clean air at a constant polarity of the applied voltage (Figure 5.25). The *CdS* sensors withstand the over-concentration of SO_2 in twice over than upper limit of the measuring range. After such overload *CdS* sensor restored its initial state in (100-300) seconds.

Presented *CdS* sensors, proposed in [74], operates at relatively low T area and allow to register the SO_2 in an inert medium, in contrast to most oxide-metal elements that are based on high-temperature thermal catalytic effect.

5.6. Sensors to Measure the CO Concentration

5.6.1. In_2O_3 Sensors for CO Control

Sensors based on In_2O_3 manufactured as one-electrode design, in which the thickness of the sensitive layer is 30-100 *mkm* and the size of the sensor sensitive element is ≈ 300 *mkm*. Power consumption of the sensor is 0.15 *W* at an operating temperature about 770 *K*. The threshold of sensitivity for In_2O_3 sensor to *CO* is 10 *ppm*, and the adsorption and desorption times in the range of (3-6) seconds, it has relatively high stability of the initial characteristics.

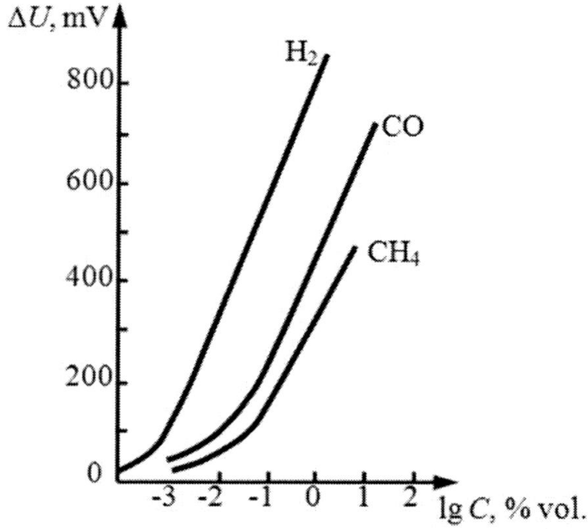

Figure 5.26. Dependences of the output sensor signal on the concentration of gases: CH_4, H_2, and *CO*.

Applying of In_2O_3 sensors allow to detect the presence in the atmosphere not only *CO*, but and other gases (Table 5.8). Results of the In_2O_3 sensors interaction with gas group are shown in Table 5.8 and Figure 5.26, its indicated to the differences in values of the threshold sensitivity and sensitivity to each of the gases. This makes it possible to reduce the negative effects of gases signal interference that simultaneously present in the measured atmosphere.

Doping on In_2O_3 sensors, for example by 10 % Bi_2O_3, makes it possible to register the *CO* in the presence of CH_4 in the atmosphere (Figure 5.27) because the maxima of In_2O_3 / Bi_2O_3 sensor adsorption sensitivity to these gases sets on at significantly different temperatures. It promotes to address an issue of the sensor selective.

Table 5.8. Parameters of the one-electrode sensors based on In_2O_3

Parameter	Value
Concentrations range in air	0-100
Output signal U_0 in air, %	1,49±0,2
Operated current, A	1,10±0,01
Maximum power consumption, W	0,18
Response time, s	3
The lower limit of detection, *ppm*	
Methane	50
Propane	10
Carbon monoxide	10
Ammonia	20
Hydrogen	1
Hydrogen sulfide	1
Maximum change in U_0 (%)	
After 2000 hours of continuous operation	2
After mechanical damage about 1000 shocks	0,2
With the acceleration of 750 m/s^2	
After the vibration frequency of 250 Hz	0,2
Weight, g	2,5
Size, mm	10x16

Simultaneous monitoring of the light oil fractions combustion process in a production environment at the thermal power plant, made by the In_2O_3 sensor (Figure 5.28, a) and industrial gas analyzer (Figure 5.28, b) show: at first – the operate properties of In_2O_3 sensor and, secondly, indicates the stability of the oil combustion process under normal conditions. Changes in the *CO* concentration at the combustion chamber provides by oxygen supply variations in the combustion zone [8].

Less stable relative to the oil is coal combustion, which fixed with comparable reliability by In_2O_3 sensor (Figure 5.29, a), and by industrial gas analyzer (Figure 5.29, b).

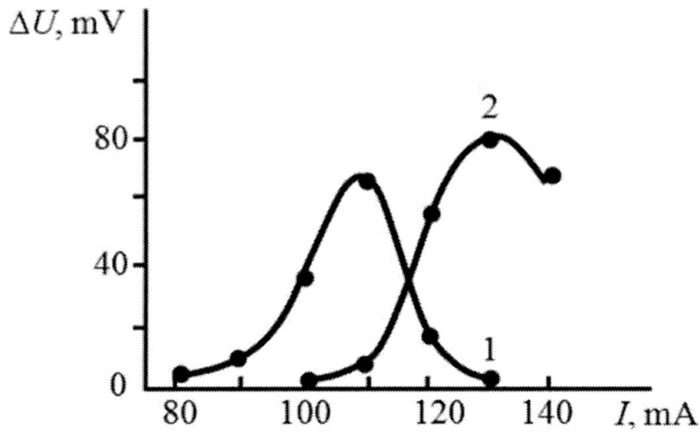

Figure 5.27. Dependence of the output signal (ΔU) value on the heating current of one-electrode In_2O_3/Bi_2O_3 sensor by exposure (1) in *CO* atmosphere with concentration of 0.1 % and (2) in the CH_4 atmosphere with concentration of 1 %.

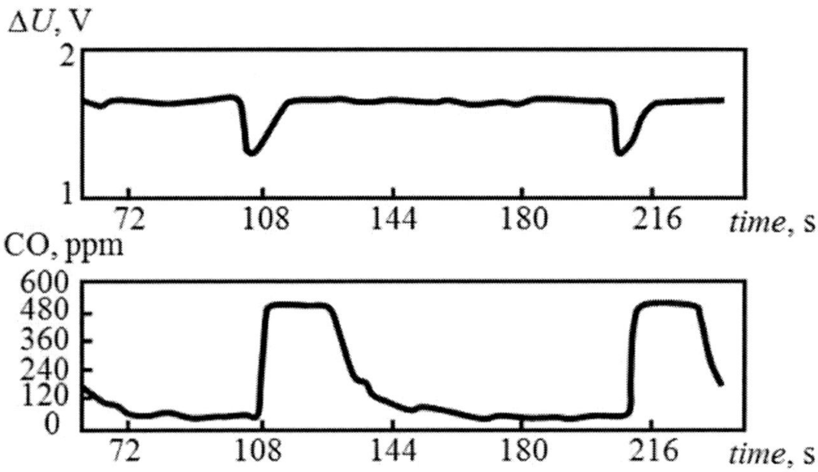

Figure 5.28. a) Changes in the In_2O_3 sensor output signal; b) changes in the *CO* concentration (*ppm*) measured by industrial gas analyzer. Measurements were carried out during the combustion of light oil fractions [8].

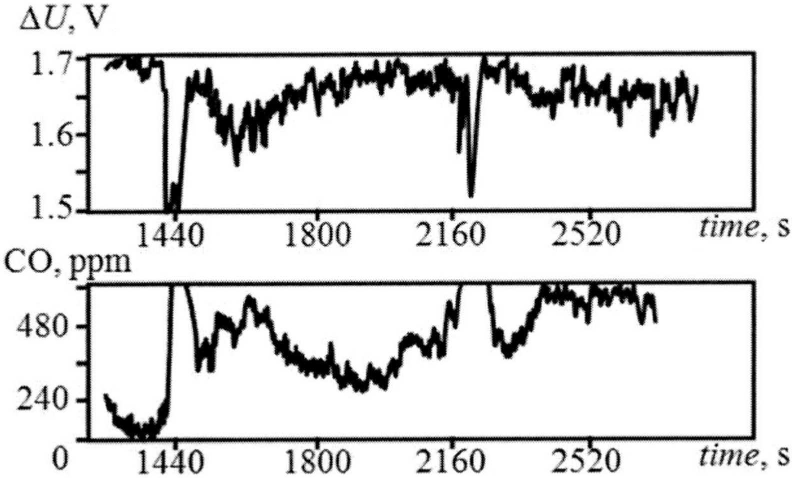

Figure 5.29. a) Changes in the In_2O_3 sensor output signal; b) changes in the *CO* concentration measured by industrial gas analyzer. Measurements were carried out during the combustion of coal [8].

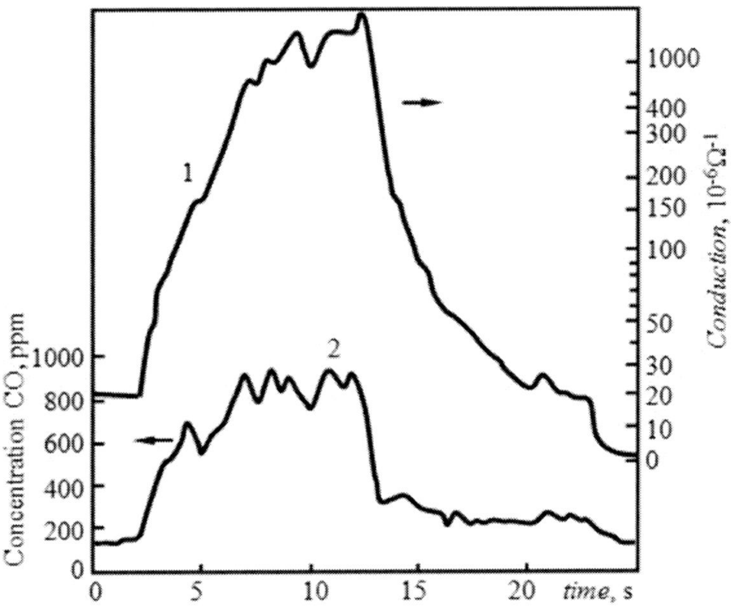

Figure 5.30. Change in conductivity of planar SnO_2 sensor for *CO* (1) in dependence on the *CO* concentration (curve 2) during combustion of light oil fractions. Measurements of *CO* concentration made by industrial gas analyzer [8].

5.6.2. SnO_2 Sensors for CO Control

Sensors based on SnO_2 fabricated by planar technology as annealed ceramics and exhibit properties that are characteristic of the *n*-type semiconductor [8].

During the tests along with SnO_2 sensors were applied the industrial infrared gas analyzer and commercial semiconductor sensor TGS.

The combustion of natural gas and light oil fractions are quite stable and accompanied by releasing a low concentration of *CO* (Figure 5.30), while the peat combustion (Figure 5.31) or wood (Figure 5.32) are less stable processes [8, 75].

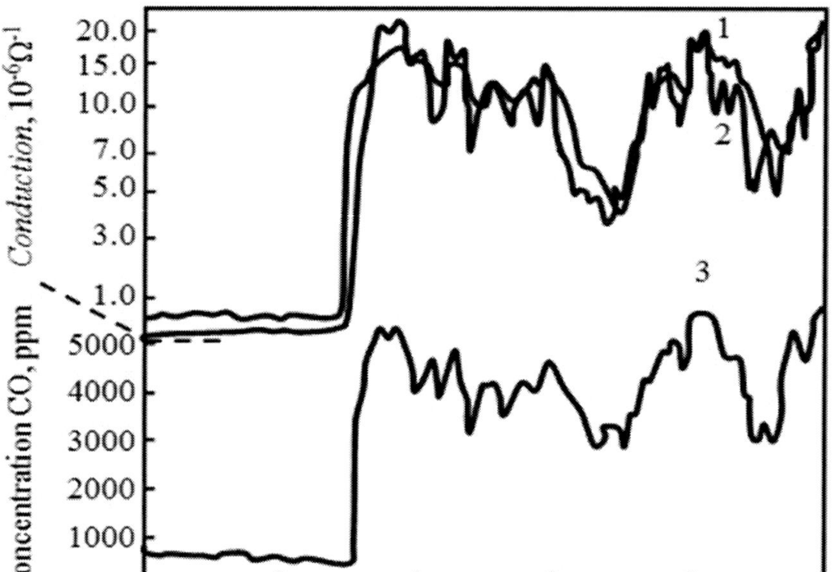

Figure 5.31. Changes in the conductivity of the two SnO_2 sensors (1, 2) and in the *CO* concentration (3) during the peat combustion. Measurements of *CO* concentration are made by independent industrial gas analyzer [8].

Increasing the number of *Sb* in the SnO_2 sensors affects their sensitivity, although the surface *Sb* cations are well known as active catalytic centers [75, 76, 77] to the *CO*. However, sensors doping by palladium (Figure 5.33) or silver (Figure 5.34) to certain concentrations are improves their quality.

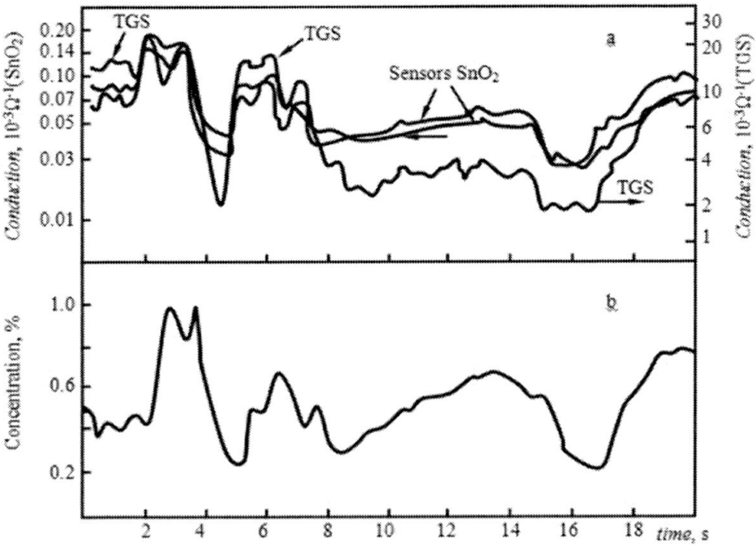

Figure 5.32. Changes in conductivity of planar SnO_2 sensors and commercial sensor TGS (a) caused by the changes in *CO* concentration during the wood combustion. Measurements of *CO* concentration are made by industrial gas analyzer (b). TGS curve in Figure 5.32(a) corresponds to the signal of sensor TGS [8].

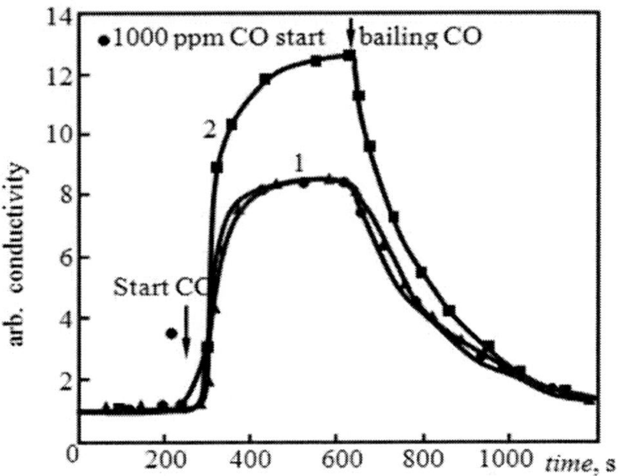

Figure 5.33. Relative change in conductivity of SnO_2 sensors (doped by *Pd* with concentration of 0.05 (1) and 0.1 (2) mole percent) in *CO* atmosphere, with concentration of 1000 *ppm* in the artificial gas mixture which simulate the exhaust gases at 820 *K*.

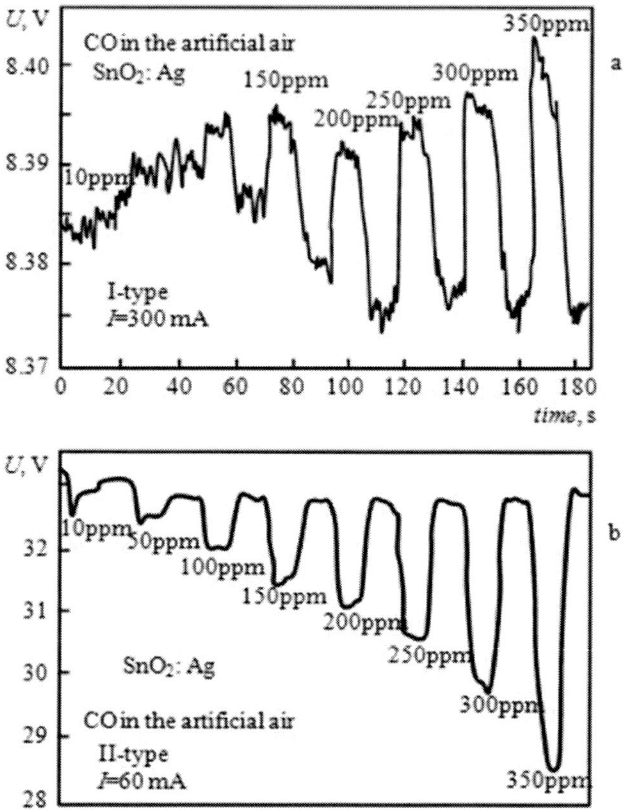

Figure 5.34. Change in sensor output signal (a) I type and (b) II type with SnO_2Ag sensitive layer upon exposure in the *CO* atmosphere of 10, 50, 100, 150, 200, 250, 300 and 350 *ppm* in synthetic air. *Ag* concentration large in sensors of type II [8].

Doping of SnO_2 or SnO_2/Sb sensors surface by platinum results to appearance a low-temperature peak of the adsorption sensitivity of Pt/SnO_2 or $Pt/SnO_2/Sb$ sensors, it located at range of $T \approx$ (350-370) *K* and has no place in the sensor without *Pt* dope (Figure 5.35).

In the structure of SnO_2 sensors the *Sn* atoms are found in the Sn^{4+} ions form in the SnO_2 phase and in the Sn^{+} ions form in the *SnO* phase , as evidenced by the two lines of $Sn3d_{5\backslash 2}$ with binding energies E = 486.7 *eV* and E_b = 485.4 *eV* (Figure 5.36). In the initial state on the Pt/SnO_2 sensor surface, in addition the actual *Pt*, contains noticeable amount of platinum in the $Pt\ (OH)_2$ (Figure 5.36, b), that could be reduced almost to zero as a result of high-temperature in a high vacuum.

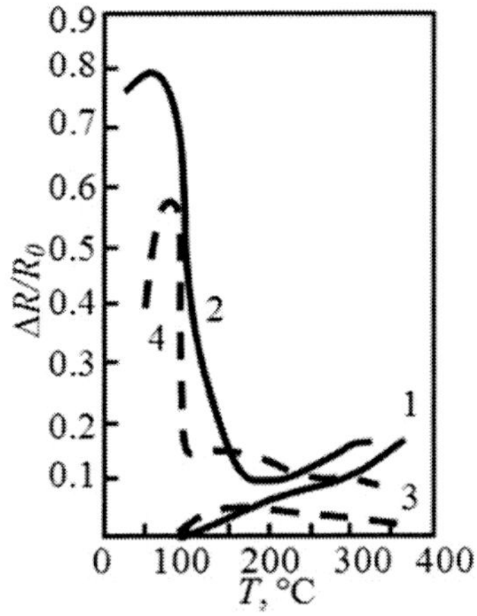

Figure 5.35. Temperature dependences of the response of sensors based on SnO_2, measured in CO atmosphere with concentration of 250 *ppm*: 1) - SnO_2, 2) - Pt/SnO_2, 3) - SnO_2: Sb, 4) – Pt / SnO_2: Sb. The response was defined as a change in the sensor electrical resistance: $f(T) = \Delta R / R$ [8].

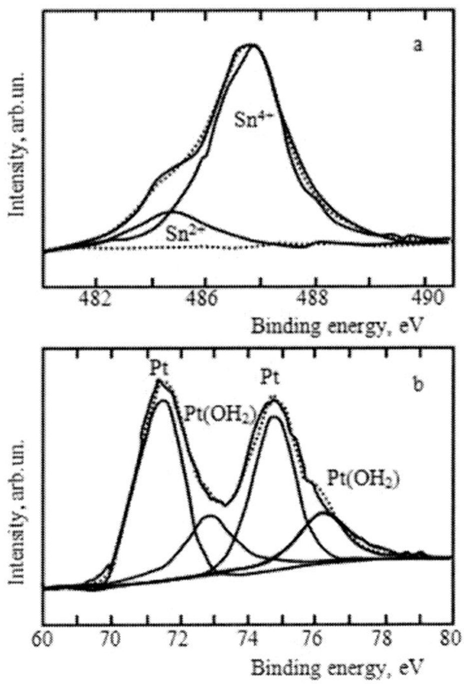

Figure 5.36. X-ray photoelectron spectra (XPS) of Pt/SnO_2 sensors: a) line of $Sn3d_{5/2}$; b) line of $Pt4f_{7/2}$ [8].

5.6.3. CdS and Sn_xWO_y Sensors for CO Control

Chemisorption of *CO* molecules on the *CdS* sensor surface with a composition close to stoichiometric leads to the increase in conductivity, i.e. the donor has the influence on the semiconductor of *n*-type conductivity (Figure 5.37, curve 1). With an increase of the cadmium consistence in the *CdS* sensor, with increasing of its number from I to III, the donor effect of *CO* is increase (curves 2 and 3, respectively). However, at low temperatures area there is the acceptor-influence effect of *CO* chemisorption on the *CdS* sensor conductivity. The higher the concentration of *Cd* in *CdS* sensor, the lower temperature transition occurs from the acceptor to donor reaction of *CO* on *CdS* sensor (curves 2 and 3, respectively) [8, 77].

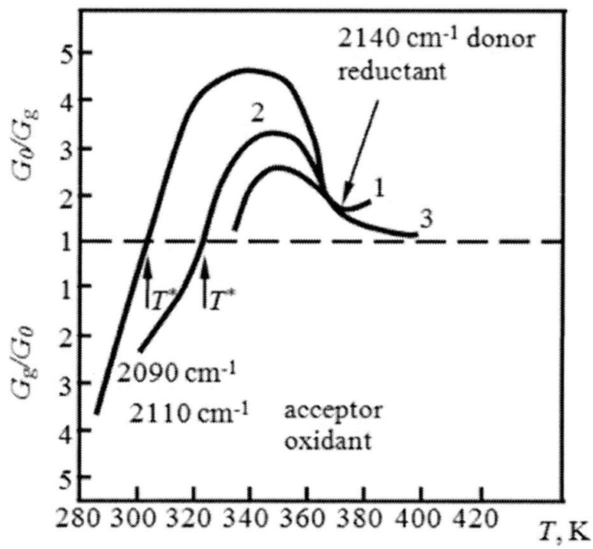

Figure 5.37. Changes in the relative conductivity G_g/G_0 (and G_0/G_g) of *CdS* sensors I, II, III types at 110 *ppm* of *CO* in nitrogen, in dependence on the temperature. The consistence of cadmium in *CdS* sensor increases with the transition from I to II and then to type III [8].

In the case of Sn_xWO_y sensors, the acceptor effect of *CO* chemisorption on its electrical conductivity takes place only under the condition that the one *Sn* atom has no more than two *W* atoms aside. Further increase of the *W* concentration in the Sn_xWO_y sensor is accompanied by the complete disappearance of the acceptor effect of *CO* chemisorption (Figure 5.38).

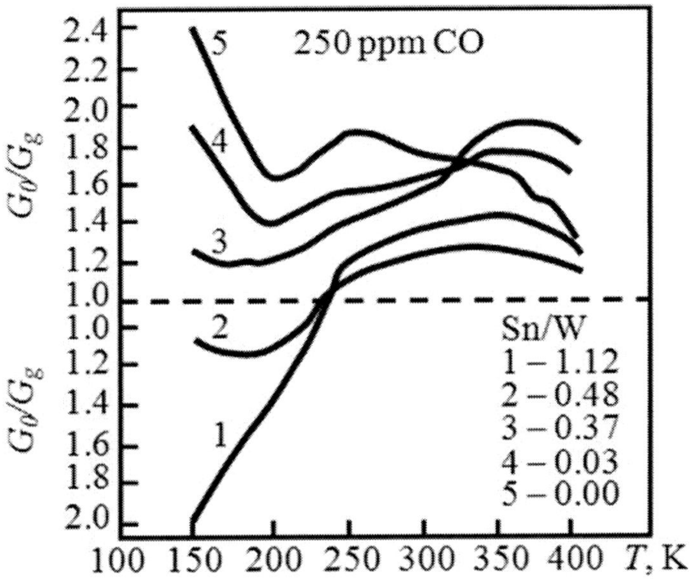

Figure 5.38. Changes in the relative conductivity G_g/G_0 (and G_0/G_g) of Sn_xWO_y sensors with different compositions at 250 *ppm* of CO in synthetic air as a function of temperature [8].

The Sn_xWO_y sensors characterized by relatively high adsorption and desorption times as at the donor and acceptor influences of CO (Figure 5.39, a), that with further increasing the operating temperature about 100 ^0C are not of noticeable decrease (Figure 5.39 b). By a much greater performance are characterized the CdS sensors, for which the adsorption and desorption times of CO as in the donor, than at the acceptor influences on the sensor is a few seconds (Figure 5.40).

A not numerous papers dedicated on the dual effects of CO on the semiconductors electrical conductivity are presented by [8, 77-80], and they do not give the proved mechanism of this phenomenon. However, the acceptor effect of the CO chemisorption on the electrical conductivity of CdS or Sn_xWO_y appears at the increasing of the metal component concentration from the group A_2 or A_4 [8, 78-80].

Acceptor effect of CO on the CdS sensor which enriched by Cd accompanied by the presence of two peaks in the IR spectrum of carbonyl adsorption forms, located at 2090 and 2110 cm^{-1} (Figure 5.41, a). Increasing of the temperature to $T \geq 340\ K$ leads to the appearance of absorption peak at 2140 cm^{-1} instead the observed at 2110 and 2090 cm^{-1}.

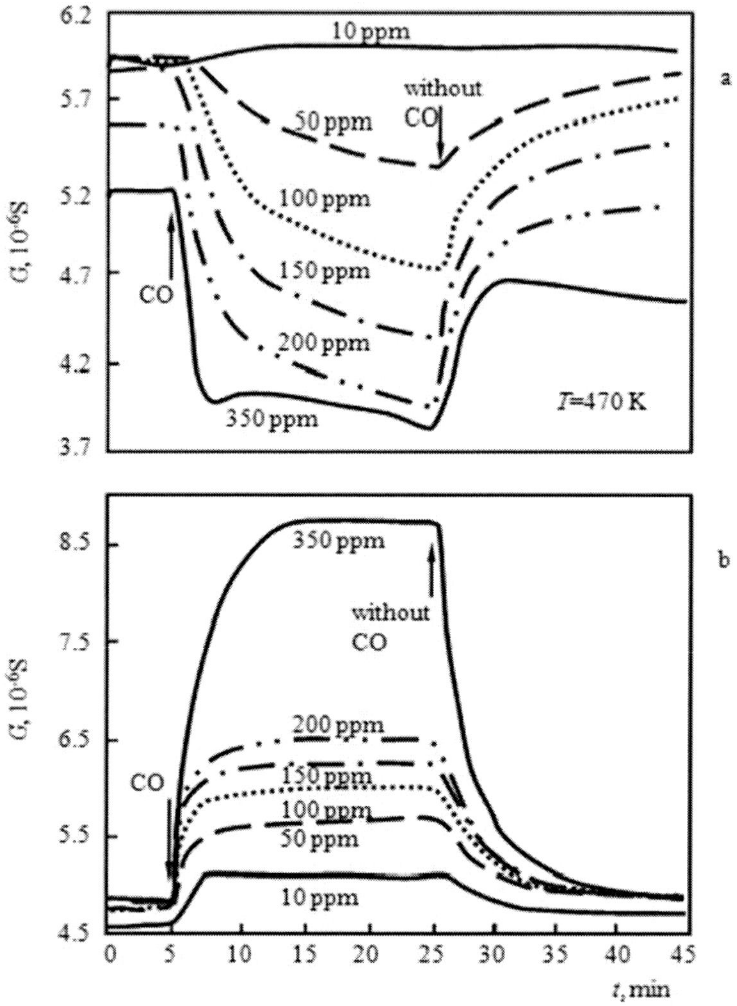

Figure 5.39. Conductivity kinetic of Sn_xWO_y sensors at the different CO concentrations in dry synthetic air at 470 K (a) and 570 K (b) [8].

Chemisorption origin of the peaks at 2110 and 2090 cm^{-1} evidenced by the fact that it both disappear after removal the CO from atmosphere. However, if the 2110 cm^{-1} peak disappears immediately after removal the CO from atmosphere, the 2090 cm^{-1} band decreases with time gradually and fully unrecorded in 15 minutes [8, 77]. Chemisorbed CO molecule on the CdS sensor surface is able to be at least in three states. Two of these states are characterized by peaks in the infrared spectrum at 2110 cm^{-1} and 2090 cm^{-1}, and accompanied by acceptor influences of the *CO* molecule on the *CdS* sensor. Third form of *CO* bond with the *CdS* sensor surface is characterized by peak in the IR spectrum at 2140 cm^{-1}.

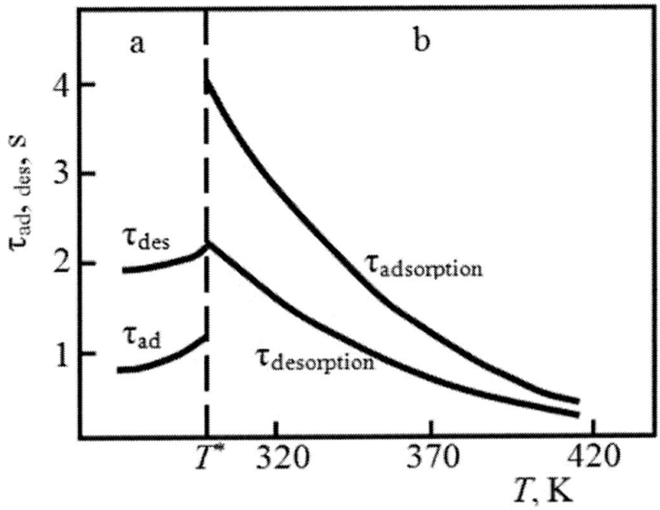

Figure 5.40. Dependence of the adsorption (τ_a) and desorption (τ_d) times of sensors at acceptor and donor interactions with *CO* on the temperature.

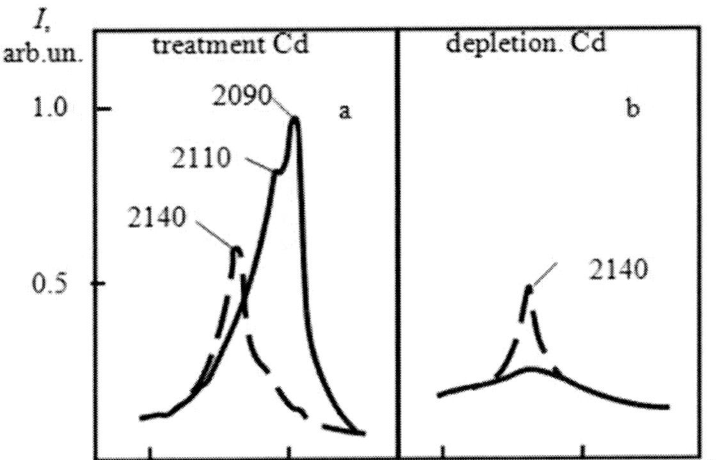

Figure 5.41. IR spectra of carbonyl adsorption forms measured after the *CO* chemisorption on the *CdS* films surface: III (a) and II (b) types at 290 *K* (solid line) and 340 *K* (dotted line).

It is known that cadmium atoms, used for doping of the compounds like *CdS* and *CdSe* types, create in the band gap at the bottom of the conduction band the small energy levels of n-type [27] and are able to locate near themselves the Fermi level. In this case, the energy levels, that created by chemisorbed *CO* molecule, in the band gap on the semiconductor surface are able to be below the Fermi level and, under certain conditions, performs the acceptor functions. Due to the temperature increases the Fermi level is shifted down from the bottom of the conduction band and created by a *CO* molecule the local energy level is able to be higher than Fermi level. Under these circumstances it performs a donor function.

5.7. Sensors for H_2S Concentration Measurement

5.7.1. One-Phase and Two-Phase Sensors for H_2S Control

Hydrogen sulphide is toxic to humans and its maximum concentration at industrial areas should not exceed about 10^{-6} % vol.

For the analysis of H_2S micro concentrations in the air applied the semiconductor sensors with operate temperature of more than 420 *K* [83-86], it provides the H_2S chemisorption in dissociative form, which is known , leads to the much more significant changes in the electrical conductivity than the physical adsorption and not dissociative chemisorption on solids surface. The dissociation of hydrogen sulfide occurs at 473 *K* and leads to sulphidation of sample surface [83-86] and decrease of its adsorption-sensitive properties.

The possibility of applying semiconductors as sensors for H_2S, which are mostly representing the metal oxides, such as SnO_2, ZnO, TiO_2, V_2O_5, CdO, NiO, Fe_2O_3 and others is under intensive studies [83-86].

Change in electrical conductivity at the H_2S adsorption, occurs due to the two processes:

1) interaction of the H^+ radicals and the surface oxygen atoms in the oxide, which is enriched the the metal surface and, consequently, an increase in the electrical conductivity value;
2) diffusion of the hydrogen atoms in the lattice depth and across the grain-crystal boundaries, which accompanied by changes in the volume conductivity.

The second mechanism leads to embedding the hydrogen atoms in the semiconductor lattice, which accompanied by an increase in the recovery time – desorption time. For the recovery time through reduction, it is recommended to apply the thin film about 1 *mkm* [84].

For determining the optimal material to the producing sensors of hydrogen sulfide, in the air atmosphere tests the thin semiconductor films: *CdSe, CdS, CdSe: In,* In_2O_3, *CdO, ZnO, PbO,* SnO_2, Cu_2O, Al_2O_3, *SiC* [29,30]. As a parameter for evaluating the sensor sensitivity to hydrogen sulfide, used the value (since the gas mixture and the measuring circuit are not changed):

$$\beta = \frac{R_{H_2S}}{R_0},$$

where R_{H_2S} - the resistance of the sensor in an hydrogen sulfide atmosphere with concentration of 10 mg/m^3, R_0 - the resistance of the sensor in the zero gas – an air. At the higher value of β is determined the optimum settings for the sensor manufacturing technology and its optimal operating temperature.

Measurements of β showed that with temperature increasing the sensors sensitivity of all shown types increases too. The SnO_2 and *ZnO* films possess the highest value of β, and its maximum value is achieved at temperatures of (550-570) *K* and (660-690) *K*, respectively (Figure 5.42, curves 1, 2).

The low sensors sensitivity at relatively low temperatures area ($T \leq 470$ *K*) is due to the physical adsorption of H_2S on the sensor surface, which accompanied by slight changes of conductivity. With temperature increasing occurs the increase of the part of chemisorption bond with the films surface (β is growing), and, finally, at enough high temperature area the H_2S molecule dissociates on the surface with formation of the H^+ and S^- ions (β reduced in result of surface sulfidation).

Proceeding from data in Tab. 5.9, should be preferred the sensors with thickness of (0.1-0.3) *mkm*, that possess relatively high values of $\beta \approx$ 2,2-2,8 and relatively small values of about $\tau_d \approx$ 80-100 *s*.

From the Tab. 5.10, it follows that the highest value of β holds the SnO_2 films, after annealing in air atmosphere with $R_0 \approx 10^7\ \Omega$. Increasing of the β with increasing of the sample resistance caused by the decreasing in the contribution of the volumetric component on electrical conductivity of the non-oxidized metal film, and also able to occur as a result of increasing the concentration of hydrogen sulfide adsorption centers on the sensor surface oxide layer. At low annealing temperatures (up to 670 *K*) part of the sensor metal film is not completely oxidized, and dependences of the dark current on the temperature of the SnO_2 sensor with $R_0 = (10^4\text{-}10^9)$ *Ohm* evidenced of the metallic character of conductivity. In high-resistance sensors ($T > 670$ *K*) observed an exponential increase of the current with temperature increasing, with activation energies of 0.1 *eV* and 0.23 *eV*.

With increasing of the annealing temperature the film composition is closer to the stoichiometric ratio. These films possess the maximum β parameter (Tab. 5.10). In this case, the chemisorption of S^- atoms complicated by the presence of oxygen atoms in the oxide semiconductor lattice, and surface sulphidation in the operating temperatures area practically does not occurs. A change in the electrical conductivity occurs due to the interaction of the semiconductor surface with hydrogen atoms.

Figure 5.41 (a, b) shows that the sensitivity of two-phase films is higher than of single-phase sensors in the areas of macro-and micro-concentrations of hydrogen sulfide. Along with the high sensitivity the two-phase oxide films possess a higher stability than one-phase films. Thus, the drift of the zero-resistance in 10 days for the $SnO_2{:}PbO$ films are 1.7 %, and for the $SnO_2{:}PbO$ films are 2.2 %. In one-phase films drift of R_0 are at (10-15) %.

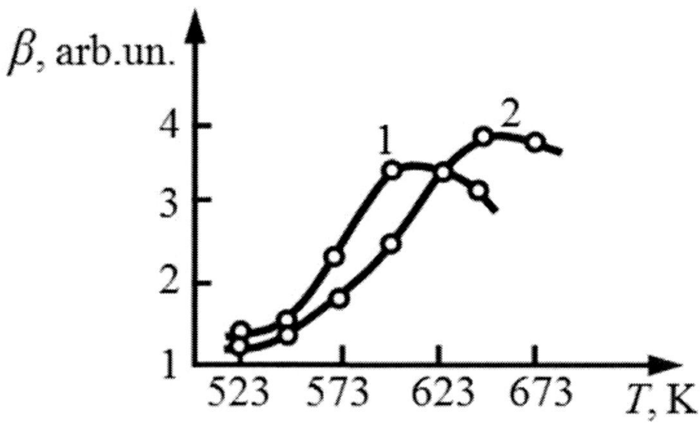

Figure 5.42. Dependence of the adsorption sensitivity to H_2S for SnO_2 (1) and ZnO (2) films on the temperature.

A two-phase oxide possesses more selective properties. Tables 5.11 and 5.12 present the results of influences the non-measured SO_2 and H_2S components studies. Featuring by a high performance on the sensitivity, selectivity and stability, the two-phase films yields to one-phase sensors on the desorption time by (1.5-2) times.

Table 5.9. Dependence of β and τ_d on the thickness of the SnO_2 sensor

d, μm	0,0,3	0,12	0,3	0,6	1,2	3,7	Notes
β	1,4	2	2,8	3,4	3,2	5	T = 700 K
τ_d, s	70	80	100	140	180	210	C = 3,5 mg/m^3

Table 5.10. Dependence of β on the resistance of sensor to tin dioxide in the zero gas atmosphere at room temperature

R, *Ohm*	10^4	10^5	10^6	10^7	10^8	10^9	Notes
β	1,05	1,1	2	2,5	2,1	2	T = 570 K C = 3,5 mg/m^3

Table 5.11. Influence of the unmeasured component of sulfur dioxide

Film type	Output signal value, mV				β
	in air	10,3 mg/m3	in air	200 ppm	
ZnO	1,8	2,8	1,8	1,7	10
SnO_2	2,85	3,6	2,85	2,7	5
ZnO : SnO_2	1	2,4	1	0,96	35
SnO_2: PbO	5,4	17,8	5,4	4,8	10

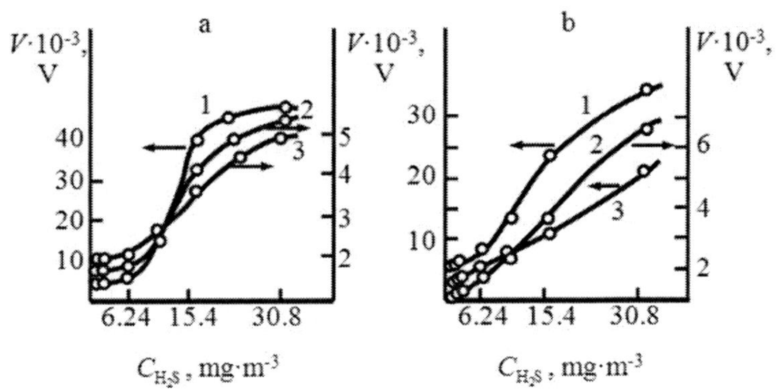

Figure 5.43. Calibration sensors characteristics: a) based on the one -phase films: *SnO_2 (1), ZnO (2), PbO* (3), and b) based on the two-phase films of *SnO_2: PbO (1), ZnO: SnO_2 (2), ZnO: PbO* (3).

One of the main sensor parameters is the basic reduced error of the δ^0 analysis, which is determined according to GOST 11.003-73 and for the best examples of SnO_2, ZnO, SnO_2: PbO, ZnO: SnO_2 there are 7.4 %, 6.8 %, 8 , 5 % and 10.5 %, respectively.

5.7.2. Doped Thick Film Sensors Based on SnO_2 for H_2S Control

The sensors' doping by the Ag atoms, as it is known, provides a significant influence on the sensors sensitivity to the H_2S [8, 84, 85]. Doped by silver the SnO_2 sensors exhibited the high sensitivity to the hydrogen sulfide in air atmosphere [86]. The output signals of SnO_2: *Ag* sensors under the influence of H_2S at different concentrations shown in Figure 5.44. The operate temperature of the sensor is about 720 *K*. At this temperature sensor possess a very high sensitivity in the field of H_2S micro-concentrations. The increase in operating temperature ($T_1 < T_2 < T_3$) leads to the increase in the output signal value of the sensor (Figure 5.45).

The sensor response time increases with increasing of the hydrogen sulfide concentration in 7 min at 2 ppm to more than 20 minutes at 3 ppm. However, after removal of the H_2S, the recovery time of the output signal value is still less than 10 minutes at the concentrations of H_2S up to 4 *ppm*.

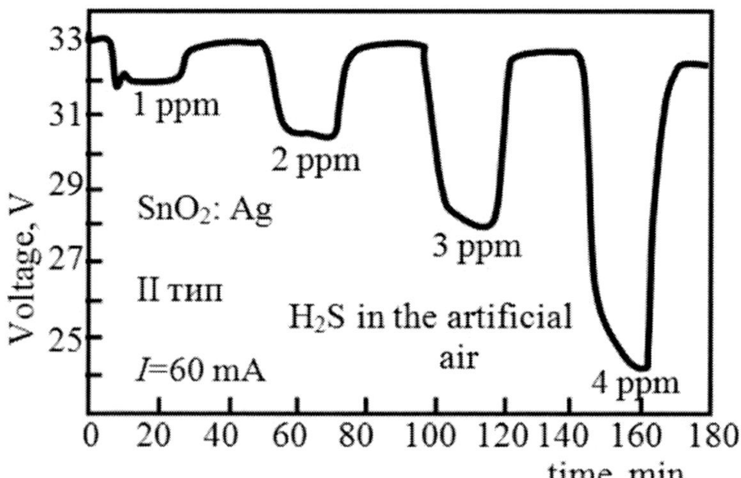

Figure 5.44. Change in the output signal value of the SnO_2:*Ag* sensor upon the exposure in the H_2S atmosphere of 1, 2, 3 and 4 *ppm* in synthetic air [8].

The SnO_2 sensors, doped by the *Sb*, also are sensitive to the interaction with the H_2S (Figure 5.46). Doping by *Sb* atoms the SnO_2 sensors, leads to increases the sensors conductivity, and, in addition, the *Sb* cations are active centers that enhance the catalytic ability of the sensors [85]. With this, the SnO_2: *Sb* sensors shows a high sensitivity to the interaction with the H_2S (Figure 5.46) in the entire researched range of its concentration [8].

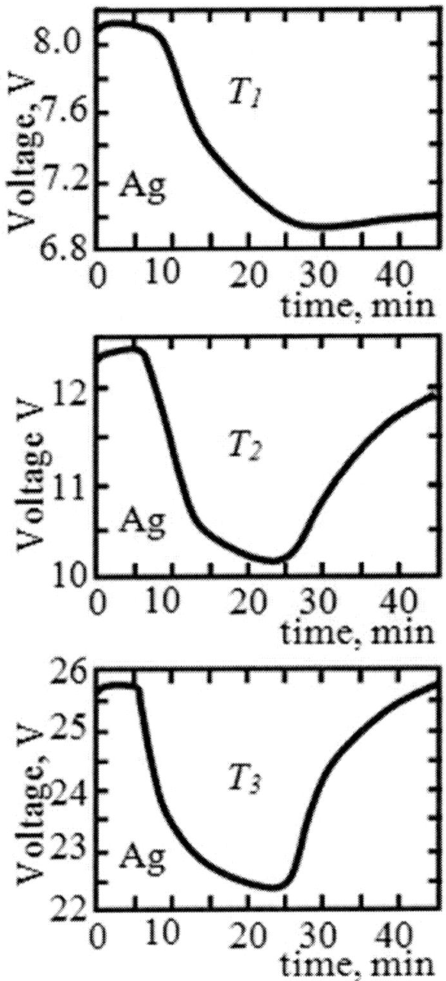

Figure 5.45. Change in the output signal value of the SnO_2: *Ag* sensor upon exposure in the H_2S atmosphere in synthetic air at the different operating temperatures: $T_1 < T_2 < T_3$ [8].

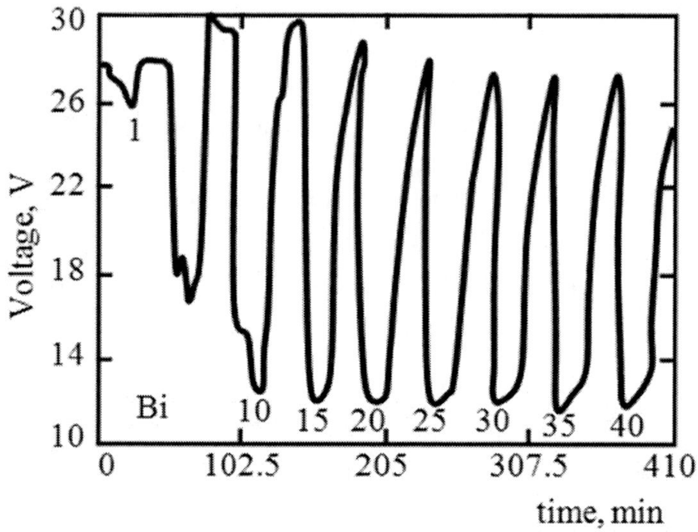

Figure 5.46. Change in the output signal value of the SnO_2: *Sb* sensors upon exposure in the H_2S atmosphere with concentration in 10-40 *ppm* [8].

5.8. THE SnO_2 SENSORS FOR MONITORING THE HYDROCARBONS CONTAINING GASES CONCENTRATIONS

5.8.1. The SnO_2 Sensors for Ethanol Control

Adsorption sensitivity of the planar thick-film SnO_2 sensors is:

$$\beta = \frac{V_{gas}}{V_{air}}$$

(V_{gas} – voltage drop on the load resistance of the measuring circuit, in the presence of ethanol in the air; V_{air} – voltage drop on the load resistance in the air atmosphere without ethanol), its increases with increasing the operating temperature of the sensor up to T = 750 K (Figure 5.47), at which adsorption sensitivity reaches the maximum and then decreases. At the same temperature the maximum is reached also for temperature dependence of the SnO_2 sensors dark current in air + ethanol atmosphere (Figure 5.48, curve 1). Presence of the ethanol in the atmosphere increases the dark current of the SnO_2 sensor and shifts the maximum in the curve *I* (*T*) at high temperatures area (Figure 5.48, curves 1 and 2).

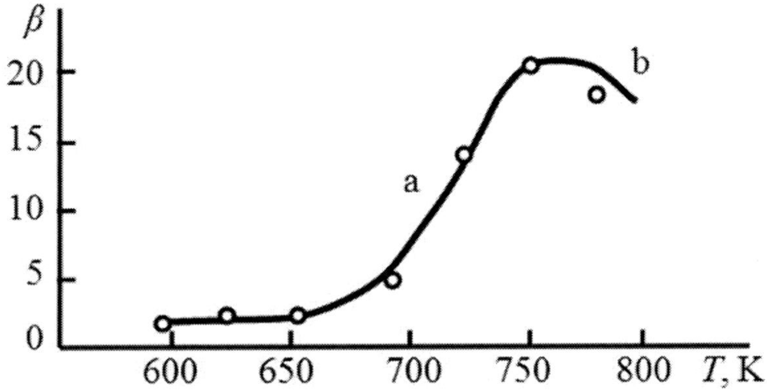

Figure 5.47. Dependence of the adsorption sensitivity of the SnO_2 sensors before annealing on the temperature: a) - thermal emission from the donor centers, b) - anomalous dependence on the temperature [8].

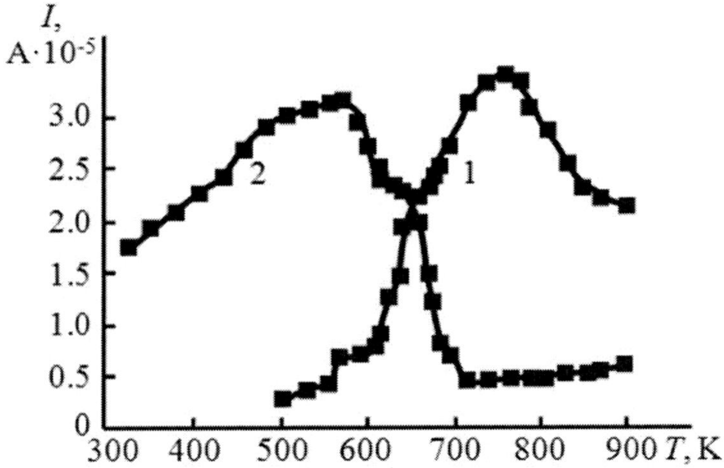

Figure 5.48. Dependence of the SnO_2 sensors dark current before (1) and after (2) annealing in air + ethanol atmosphere on the temperature.

The minimum value of adsorption time is achieved at the same temperature in 750 *K* (Figure 5.49, curve 1), as the maximum of β (Figure 5.47), and desorption time τ_d decreases monotonically with increasing the operating temperature (Figure 5.49, curve 2). After annealing of the SnO_2 sensors in argon atmosphere at 870 K for 1 hour, its sensitivity β significantly reduced, but optimal operating temperature (Figure 5.50) decreases for 50 K (up to T = 700 K), and power consumption of the sensor is reduced respectively.

After heat treatment in the argon atmosphere the adsorption time τ_a decreases (Figure 5.51, curve 2). By the exponential dependence of τ_d on temperature it is found that the binding energy of the ethanol molecule with the SnO_2 sensor surface is 0.23 eV [77], which well agrees with the values of 0.22 ... 0.26 eV that obtained in [8].

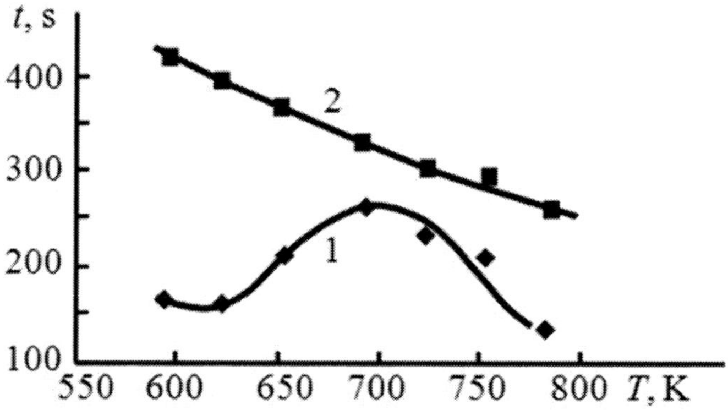

Figure 5.49. Dependence of the adsorption τ_a (1) and desorption τ_d (2) times of the SnO_2 sensor on the temperature, measured before its annealing [8].

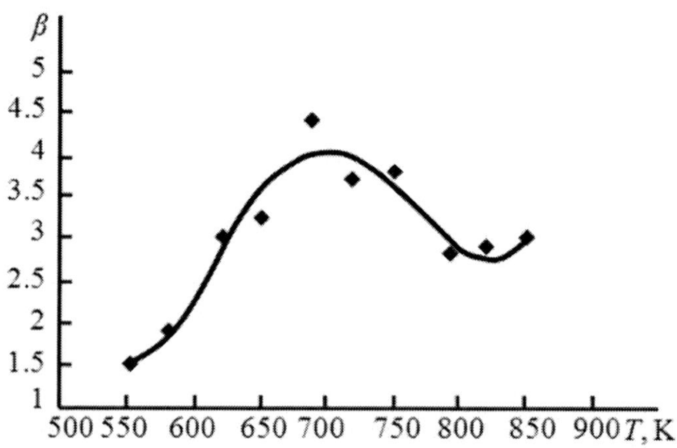

Figure 5.50. Dependence of β of the SnO_2 sensor on temperature after its annealing in the argon atmosphere at T = 870 K for 1 hour.

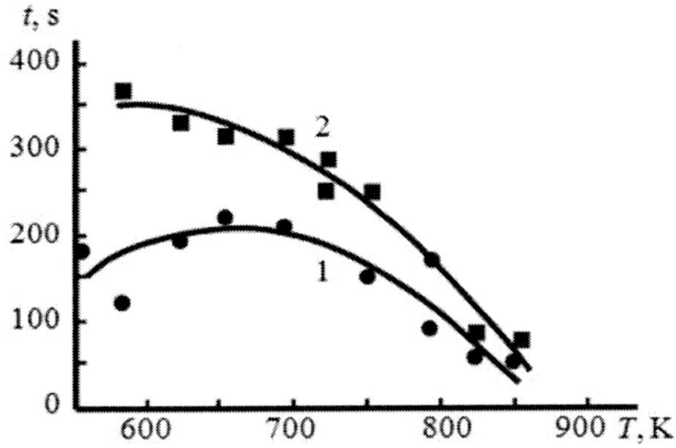

Figure 5.51. Dependence of the adsorption (1) and desorption (2) times on the temperature after annealing in the argon atmosphere at $T = 870\ K$ for 1 hour.

5.8.2. The In_2O_3 sensors for Monitoring the CH_4 Presence in the Atmosphere

Study of the In_2O_3 single-molecular sensors sensitivity to the methane (1 % CH_4 in gas mixture of 80 % N_2 + 20 % O_2) found that the maximum of the sensor sensitivity to CH_4 observed at operating temperatures (720-770 K, $I \approx$ 140 mA) (Figure 5.52), and its operation differ of a relatively high stability (Figure 5.53).

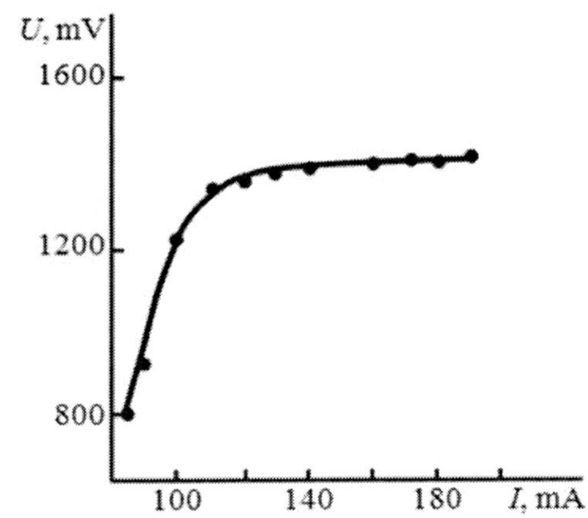

Figure 5.52. Dependence of the output signal (U) of one-electrode In_2O_3 sensor in the gas mixture (1% of CH_4, 80% of N_2 and 20 % of O_2) on the operating current value.

Adding platinum in the *In_2O_3* sensor structure, reduces by 80 *K* its operating temperature for the *CH_4* detecting [8]. By XPS method founds that the composition of *In_2O_3* sensors contains *In_2O* (E_b = 443, 2 *eV*), *InO* (E_b = 443, 7 *eV*), *In_2O_3* (E_b = 444, 2 *eV*) and *In* (*OH_3*) (E_b = 445, 0 *eV*). After the doping by platinum (5% by weight), it contains the Pt atoms in different valence states (Pt^0, Pt^+, Pt^{2+}). The elemental composition of the sensors is shown in Table 5.12 [8]. As a result of the doping by platinum, the *In_2O_3* sensor operating temperature decreased by 80 *K*, as in the case of doping it by *Pd* atoms [8] and equals to about 720 *K* (Figure 5.54).

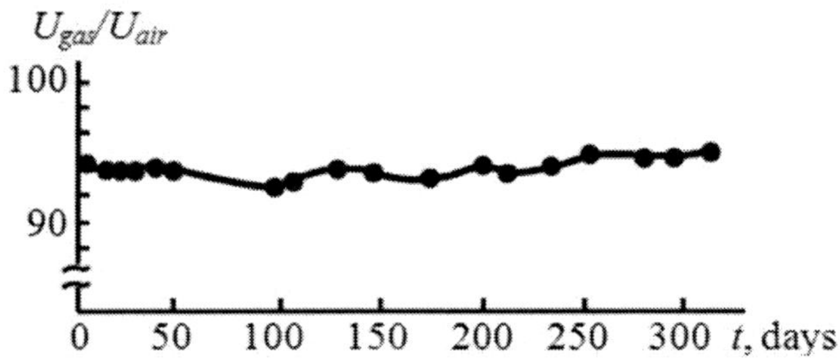

Figure 5.53. Influence of the operating time of the one-electrode *In_2O_3* sensor on its sensitivity upon exposure in the atmosphere of 80% *N_2* and 20% *O_2*, in the presence of methane with concentration of 0.1%.

Table 5.12. The elemental composition of the *In_2O_3* and *Pt/In_2O_3* samples surface, obtained by the XPS method

	In_2O_3	Pt — In_2O_3			
T, K	O/In	In, %	Pt,%	O,%	
20	1,70	34,6	2,0	63,4	
100	1,68	34,6	1,6	63,8	
150	1,65	34,9	1,4	63,6	
200	1,60	37,0	1,4	61,2	
250	1,55	37,5	1,5	61,0	
300	1,53	37,8	1,7	60,5	
350	1,52	38,0	2,0	60,0	
400	1,52	38,4	1,9	59,7	
450	1,51	39,0	2,0	59,0	
500	1,51	39,0	2,0	59,0	
550	1,49	39,0	2,5	58,5	O/In = 1,5
600	1,50	39,2	2,8	58,0	O/In < 1,5

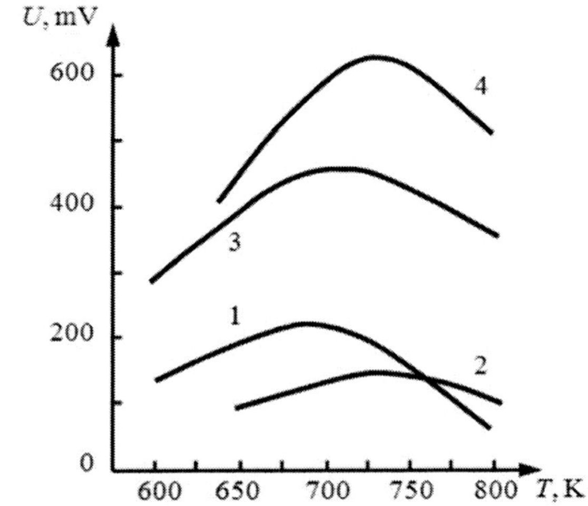

Figure 5.54. Dependence of the In_2O_3 (2,4) and Pt/In_2O_3 (1,3) sensors sensitivity to CH_4 on the temperature, measured in the atmosphere which contains methane with concentration of 1 % and 100 % (3,4).

The output signal of the In_2O_3 and Pt/In_2O_3 sensors consists of two components: a) in case of Pt/In_2O_3 sensors, the signal caused by the methane adsorption up to $T < 643$ K, and at higher T its value determined by catalytic oxidation of methane on the sensor surface; b) for In_2O_3 sensors the output signal up to $T < 723$ K caused by the adsorption of methane, and at higher T determined by the process of CH_4 catalytic oxidation on the sensor surface [8].

The sensor sensitivity decrease with increasing the intensity of the CH_4 catalytic oxidation indicates on a significant influence of the chemisorption process on its sensitivity. This is confirmed by the fact, that the sensors keep operability in the atmosphere of 100 % CH_4 [8].

5.9. Sensors Based on Porous Silicon for Ammonia Control

5.9.1. Capacitive por-Si Sensor for NH_3 Control

As a result of the special treatment of the *n*-type silicon crystals, on its surface formed the macro-porous structure of por-Si [28, 77]. Capacitance value of these structures is a few *pF*. Upon the NH_3 adsorption, capacity of these structures does not change and it are not used as the NH_3 sensors.

At *por-Si* substrates formed the micro-porous structure of *por-Si* [30], which capacity reaches ten *nF* and increases about 10 times due to the NH_3 adsorption [29, 77].

Compared to the change of the adsorption sensitivities β_n and β_μ, the variations of adsorption sensitivity with capacity β_c of similar sensors are the highest and define the changes in the total β value [29]. Therefore, there are attributed to the sensors of capacitive type.

Discussing sensors possess the internal electromotive force U_a at porosity exceeding 12%, the value of which reaches its maximum at porosity of (43-45) %. This is how the maximum influence of NH_3 adsorption was achieved on U_a value.

The value of adsorption sensitivity of the micro-porous *p-Si* depends on the concentration of NH_3 in the dark (Figure 5.55, curve 1) and at illuminated by the white light (curve 2), increasing of which intensity value decreases the value of β and shifts its maximum to higher concentrations area (curve 3).

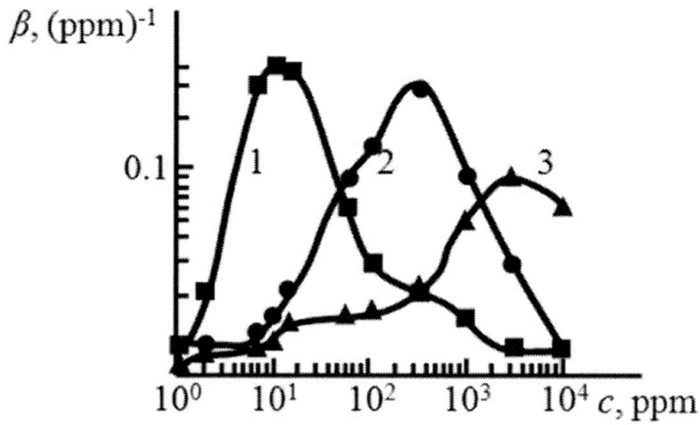

Figure 5.55. Dependences of the sensor adsorption sensitivity to ammonia, measured in the dark (1), by illumination with light intensity of 65 lux (2) and intensity of 95 lux (3).

In the studied sequence of the gases (O_2, *NO*, NO_2, *CO*, CO_2, CH_4) only ammonia shows the adsorption effect on the micro-porous *p-Si* sensor. Effect of oxidizing gases does not occurs, due to the presence the fluorine atoms on the sensor surface, and H_2 and CH_4 does not demonstrate its influence due to high hydrogen content that embedded in the near-surface area at the manufacturing process [77].

The physical reason of the changes in the electromotive force magnitude under the NH_3 adsorption is the physical adsorption of dipolar ammonia molecules [29, 36, 77]. Dipole moment of NH_3 molecule is 1.46 D, it is comparable to the dipole moment in the sensors structure, which formed by the size heterogeneity and pore distribution in it [29, 36, 77].

5.9.2. Photo-Luminescent Sensor Based on Nano-Porous p-Si for NH3 Control

Nano-porous *p-Si* sensors obtained by electrochemical etching of the single-crystal silicon, showed the photoluminescence (E_{max} = 1.7 *eV*, λ_{exc} = 441.2 *nm*) with porosity of 55 to 75 % (Figure 5.56) [29, 77].

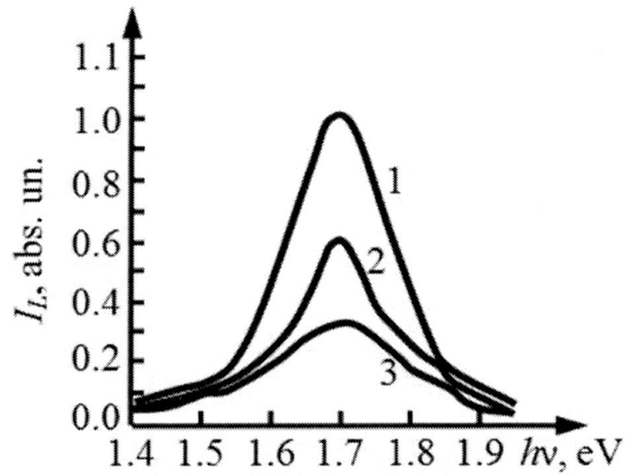

Figure 5.56. Photo-luminescence spectra of the micro-porous silicon, measured in high purity nitrogen (1), and ammonia concentration of 110 *ppm* (2) and 490 *ppm* (3) in high purity nitrogen.

Adsorption of NH_3 molecules reduces the PL intensity *I*, so the sensitive threshold is 7 *ppm* of NH_3 (Figure 5.57, curve 1), and at area of concentrations greater than 600 *ppm* of NH_3 the photoluminescent sensors lose sensitivity to an increase in the ammonia concentration. Using the formula for the adsorption sensitivity β determination from [29], in [29, 36] found that the maximum value of the photoluminescence adsorption sensitivity β for micro-porous *p-Si* sensors corresponds to the NH_3 concentration of 26 *ppm* (Figure 5.57, curve 2).

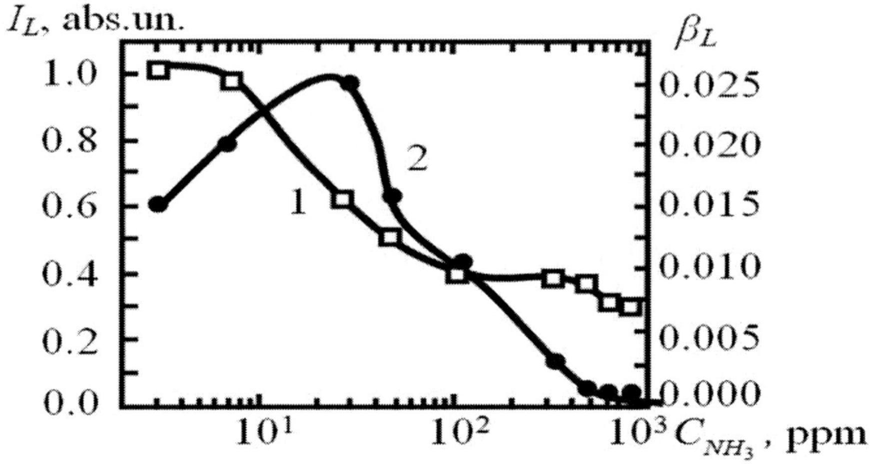

Figure 5.57. Dependences of the photoluminescence intensity (1) and value of β_L (2) on the ammonia concentration.

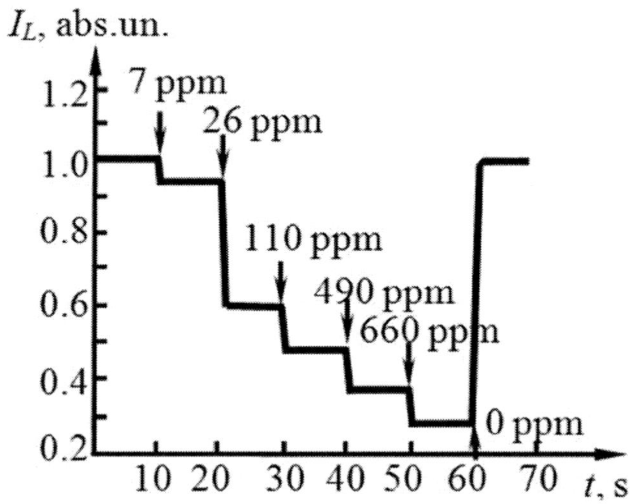

Figure 5.58. Kinetics of the change in the luminescence intensity of nano-porous *p-Si* sensor at the sequential increasing in the ammonia concentration in the test chamber up to 670 *ppm*, and then blown by the high purity nitrogen.

Photo-luminescent nano-porous *p-Si* sensor operates at room temperature and characterized by the adsorption τ_a and desorption τ_d times less than the time required to change the NH_3 concentration (1s) in the test chamber (Figure 5.58). After exposure the maximum NH_3 concentration (600 *ppm*) of the used and purge the chamber by nitrogen, the initial value of the photoluminescence intensity is restored (Figure 5.58) due to the fact that the NH_3 molecules are physically adsorbed on the sensors surface [29, 77].

5.10. DEGRADATION OF SEMICONDUCTOR FILM GAS SENSORS

Degradation of the gas sensors parameters while it operates is a negative process, which developments try to avoid in every possible ways by using different technological approaches and solutions. It is difficult to find works, in which the all possible degradation mechanisms of the gas sensors are considered comprehensively along with the factors, which influence it. Difficulties in carrying out such complex researches are caused by the fact, that in the dependence on the sensor material, operating temperature, gas composition and pressure of the gas atmosphere, it is able to develop simultaneously and consistently, excepts the adsorption and desorption of the gas, the structural changes: oxidation, reduction, diffusion processes, exciting as a surface since a volume of the sensitive sensor material [27, 36].

Obviously, the complex of such investigations can be carried out sequentially for different sensor-gas pairs.

In most cases, for determining the oxygen concentration applied the sensors temperature more than 500 *K*. In this case, for example, as in the case of zinc oxide [90, 91], and in the case of cadmium selenide films, there is an irreversible increase in resistance value and a decrease in the adsorption sensitivity (sensor degradation). Analysis of the adsorption - diffusion and the redox processes on the cadmium selenide films surface showed, that the degradation of the adsorption sensitivity and ρ linked to several sequentially and simultaneously developing processes. Increasing of the temperature more than 390 *K* stimulates the active decomposition of the adsorbed molecules O_2^- into atoms. This results in a rapid increase of the ρ due to the change of the form of oxygen chemisorption on O_2^- to O^-. A further increasing in the temperature value is accompanied by the diffusion of atomic oxygen in the volume of the semiconductor film on grain-crystals boundaries. The diffused oxygen increases the specific resistivity of the film, due to the growth of inter-crystalline barriers, caused by the localization of the conductive electrons at the oxygen centers [29, 36, 77].

At sufficiently high temperatures develops a redox reaction of the semiconductor surface layer with the oxygen, in this process the cadmium atoms (the adsorption centers) enter into a bond with oxygen and formed the cadmium oxide, then lose its activity as the adsorption centers and donors [29, 36, 77].

The mechanism that responsible for the degradation of the sensors based on oxide compounds such as zinc oxide [90, 91], considered the diffusion of the non-stoichiometric metal atoms on the surface, where its formed a chemical bond with the oxygen and completing the oxide lattice.

In the case of long-term sensors operation the degradation of its parameters be able to occur under the influence of its structure recrystallization. Sensors, for example, based on compounds of A_2B_6, possess a mixed phase structure with a predominance of the hexagonal phase on the cubic phase. Due to the recrystallization, the number of cubic metastable phase decreases. Simultaneously, being the ordering of the sensor material volume, accompanied by an increase of ρ, decreasing in the donor and adsorption centers concentration, and consequently by a decrease in the adsorption sensitivity value [29, 36, 77].

Simultaneously, but with different rates, the development of the chemisorption, oxygen diffusion and oxidation-reduction of its interaction with the cadmium selenide film, leading to the degradation of its adsorption sensitivity, is clearly seen in studies of the sensor resistance stability in time.

At $T = 520\ K$ the sharp increasing in the resistance value of all sensors, associated with the oxygen chemisorption, observed during the first minutes in the future continues at a slower pace, which has slowed down after the diffusion front reaches the substrate (Figure 5.59, curves 1-6) . Depending on the sensor material thickness and the temperature of its pre-annealing, the saturation of the sensor film by the oxygen and stabilization of its resistance occurs in (50-100) hours. During this period completed the structural changes. Further, it is beginning to affect the oxidation process of the sensor material, leads to the formation of cadmium oxide and, consequently, to a decrease in sensor resistance value. Complete conversion of the cadmium selenide sensor film in cadmium oxide is faster, than smaller its thickness and higher the pre-annealing temperature. The *CdSe* film sensor resistance of 0,4 *mkm* in thickness, pre-annealed at $T = 630\ K$, stabilizes after 1400 hours, while the resistance of the non-annealed film, with the same thickness, reaches a steady value after 2500 hours (Figure 5.59, curves 5 and 4).

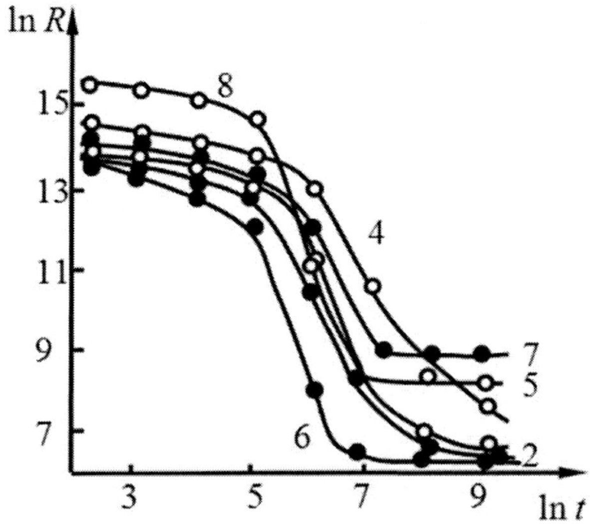

Figure 5.59. Test results of the thickness (*CdSe* sensors - 2, 4, 8) and pre-treatment temperature influence (*CdSe* sensors - 5, 6, 7, 8) on the time stability of the *CdSe* sensors resistance at operating temperature of 520 *K*. The sensors thickness are 2 - 1,26 *mkm*, 4 - 0,675 *mkm*, 5 - 8 - 0.4 *mkm*. Preliminary heat treatment temperature: 2, 4, 8 – non-annealed, 5 - 630 *K*, 6 - 520 *K*, 7 - 400 *K*.

Increasing in the thickness of the non-annealed *CdSe* sensor film to 1.26 *mkm*, is accompanied by the increase of the time value, which is required for the resistance stabilization, up to 3500 hours (curve 6). In any case, the resistance stabilization signifies the completion the complete oxidation of the *CdSe* sensor film to the cadmium oxide, that is verified by diffractometry method (after 3000 hours on the diffraction patterns of all samples are found only the cadmium oxide lines) and by the film color (the initial film of cadmium selenide is black), after oxidation it becomes colorless, which is characteristic of cadmium oxide [28, 30]).

These facts shows, that the oxidation process of cadmium selenide sensors films occurs not only due to the metal atoms diffusion from the volume to the crystallites surface, but due to it follows from [28, 30], and due to result of oxidation-reduction reactions.

Due to the pre-annealing the *CdSe* sensors thin film becomes much more stable. To the degradation reduce also contributes the operating temperature decreasing to 360 *K*. During more than in 200 hours at this temperature observed the insignificant changes in the resistance value of these films

(Figure 5.60, curves *3-5*). However, thicker *CdSe* sensors films are less stable (curves *6* and *7*).

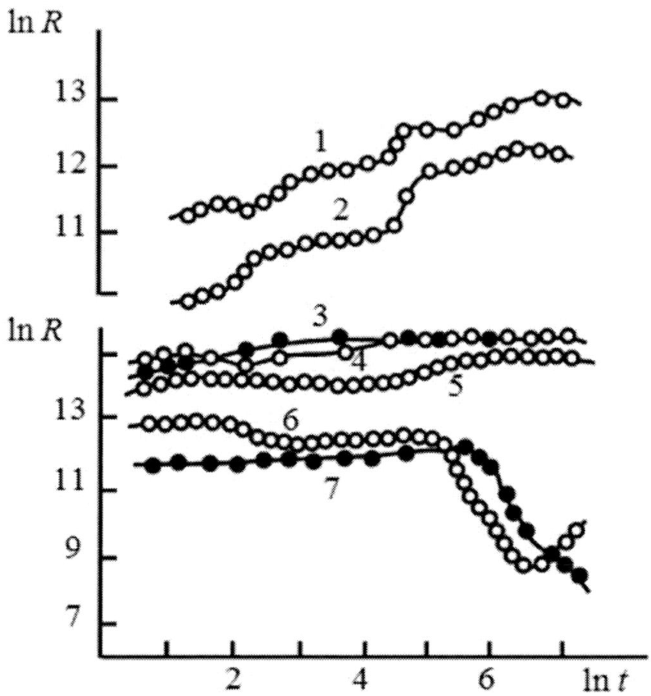

Figure 5.60. Test results of the stability in time of the *CdSe* sensor working parameters (the resistance), shown in constantly activated mode (*1, 2*) and pulse mode (*3-7*) of the sensor. The operating temperature of the *CdSe* sensor is 360 *K*, the pre-treatment temperature is 700 *K*. The *CdSe* sensor thickness is 0.3 *mkm* (*1-5*), 0.6 *mkm* (*6*), 1.2 *mkm* (*7*).

As follows from the curves *1* and *2* (Figure 5.60), the stability of the film resistance depends strongly on the measurements conditions of this parameter. The films are stable, if connected in the measuring circuit only during the registration of the resistance value, i.e. in a pulsed mode. At the full-time operating of the measurement circuit, it is observed the significant degradation of the film resistance, which, as follows from [28, 30], is related to the diffusion of oxygen ion-atoms and ion-molecules on the crystallites surface from the cathode to the anode. Thus, it is obvious, that the increase of the sensors parameters stability such as *CdSe* type, are able to be achieved by pre-heat treatment and by decreasing of the operating temperature. However, in the latter case, it sharply deteriorates the sensitivity and inertia of the sensors.

Effective increasing of the adsorption centers concentration on the sensors surface able to be achieved by doping of its surface by the donor metals atoms [29, 30]. As a result, due to the surface doping to the optimum concentration of the adsorption centers, there is a decrease in the binding energy of the chemisorbed oxygen molecule from 0.8 *eV* to 0.65 *eV*. This enables to decrease the operating temperature of the *CdSe* sensor to the (340-360) *K*, without affecting its sensitivity and performance. This significantly slows down the degradation of *CdSe* sensor, due to the formation and diffusion of the oxygen atoms. Because there are complicated the forming of the oxygen atoms, that possess a highly reactive ability, so it is not observed the formation of cadmium oxide. The intensity of structural changes also reduced, due to a decrease in operating temperature and the pre-annealing treatment.

Decreasing of the sensor operating temperature from 520 *K* to (350-355) *K* at saving the high sensitivity and performance being possible, due to the doping of the cadmium selenide films surface by indium atoms [29, 30]. During the doping of the surface, significantly reduced or, at the indium concentration of $\geq 10^{20}$ m^{-2}, eliminated the surface potential barrier. In the latter case, the adsorption limits only by the barrier of chemical nature. Liquidation of the biographical surface potential is possible due to the delocalization or compensation of the initial negative charge, occurring in the process of accumulation the positive charge of the ionized indium atoms on the surface. Simultaneously decreases the τ_d, because decreases the binding energy between the chemisorbed molecule O_2^- and adsorption center, which can be regarded as a barrier that limiting the desorption process. Its decreases from 0.8 *eV* to 0.65 *eV* is a result of the dipole interaction between adsorption complexes and the result of the influence on the adsorption complex binding energy the crystal field of the indium clusters [29, 30, 31].

5.11. Sensitivity and Selectivity of the Semiconductor Film Gas Sensors

5.11.1. Adsorption Sensitivity of the Gas Sensors

Determination of the adsorption sensitivity β of the gas sensors [30], as follows from the analysis of formula (2.2) in [29], makes it possible to consider and theoretically predict the influence of the electronic parameters on β of these devices. One of the technologically driven and important parameters in this formula is the thickness *d* of the sensor material. Increasing in the *d* of the sensor material leads to decrease of the influence of surface processes on the current transport in the sensor and to decrease of its value β [29, 30].

Experimentally, in most cases, the dependence of β (*d*) is more complicated. This is caused by the fact, that at various technological stages of the film formation process (the layer of the crystal) of the gas sensor (nucleation, the formation of the transition layer, continuous film, self-doping structure by the intrinsic defects, structure re-crystallization [29, 30]), the chemical nature and concentration of adsorption centers on the its surface is change. Due to such variations, changes not only the value of the parameter β, but its sign, i.e. its enable to change the reaction mechanism on the surface, for example, from the donor to the acceptor mechanism [8, 29 - 31].

During the derivation of (2.2) in [29], it does not take into account, as observed in [29, 30], the irregular distribution of the properties, for example, up to the film thickness of the *CdS* and *CdSe* sensors. As the major donors at the same time can be the adsorption centers, the increasing in the concentration of selenium and sulfur vacancies at the initial stages of growing is able to lead to the increasing of β, as well as the "leakage" of the cadmium atoms on the surface due to the increase in N_{ds}. If in this case considered that during the growth of *CdSe* sensor film increases the G_s contribution to the overall conductivity, it is obviously that the experimentally observed increasing in β with increasing the *d* (Figure 5.61) cannot be fully described by the formula and caused by the described in [29, 30] process of self-doping the *CdSe* sensor film. There also shown that after the process of self-doping begins the developing of self-compensation, that ultimately leads to the re-crystallization of the *CdSe* sensor film and, consequently, to a decrease in N_{ds}. After a sharp decrease of G_s the main factor, that influencing on the β, is inversely proportional dependence σ_{sc}^{0} on *d*, and so the β is decreases. Thus, only at sufficiently large *d* values, where the cadmium selenide sensor film is represents the recrystallized stable mono-phase system, should expect the correspondence between the calculated and experimental dependences of β (*d*), that is indeed right (Figure 5.61 and 5.62).

Only at $T > 450\ K$ observed the predicted in [30] the β decreasing with the increase of *T*.

Are obtained the experimental dependences of β (*T*) at area of low operating temperature of the sensor (Figure 5.63, 5.64) are not correspond to the calculated data.

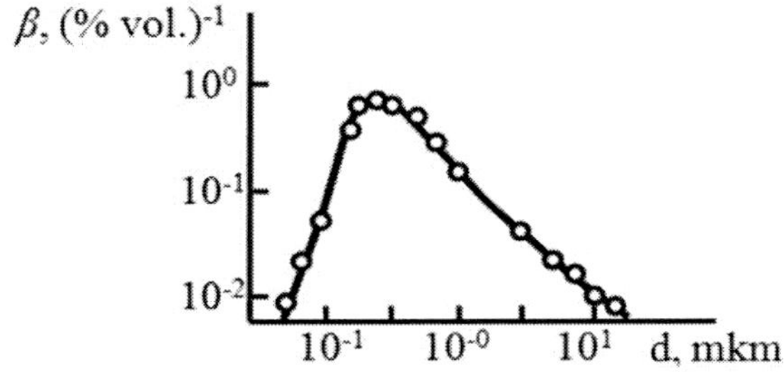

Figure 5.61. Experimental dependence of the adsorption sensitivity on the cadmium selenide sensor film thickness. Measuring temperature is 420 *K*.

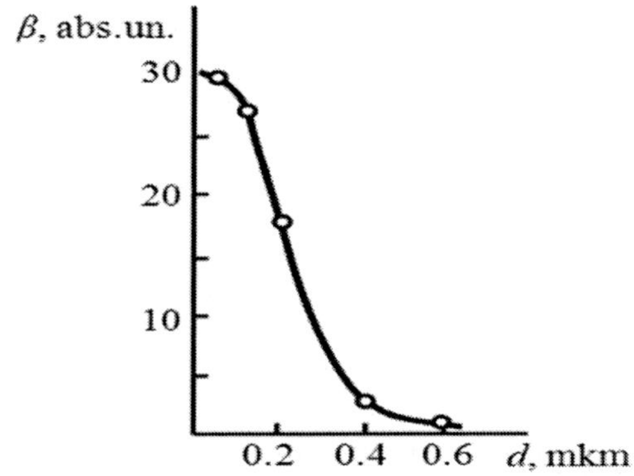

Figure 5.62. Dependence of the sensitivity to the sulfur dioxide on the *CdS* sensor film thickness.

Of course, the considering definition of β cannot be useful in the analysis of experimental dependence of the β value on technological parameter such as substrate temperature, on which condensed the sensitive layer of the sensor (Figure 5.65). On the other hand, the theory is good description the experimental dependences of the adsorption and desorption times of the gas particles [92] (Figure 5.66, 5.67) and in some cases allow to determine the energy parameters of the adsorption and desorption processes [29, 30].

Researches of the contact potential residual and the kinetics of electron-molecular processes [30, 31] shown, that the increasing in the sensor layer temperature stimulates the transition from the molecular to the atomic form of oxygen chemisorption. Occurs at the same time the sharp increase in the E_{ta} value (from 0.75 eV to 1.2 eV) and value of N_s (from 5×10^{18} to 9×10^{18} m^{-2}), according to [30], should leads to an increase in value β. With increasing in T

growth the rate of O_2 dissociation, the concentration of the chemisorbed atoms tends to the limit value. In result of the removal the Fermi level, with an increase in *T*, from the conduction band increases the probability of releasing the trapped electron on the oxygen energy level of the adsorption origin. These processes slow down the growth of β with T increasing and, in the end, as predicted in [30], leads to a decrease in β.

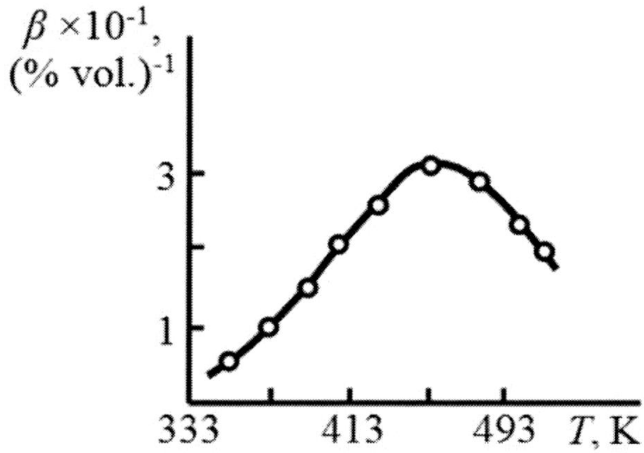

Figure 5.63. Dependence of the adsorption sensitivity to the oxygen on the temperature of the cadmium selenide film, measured experimentally.

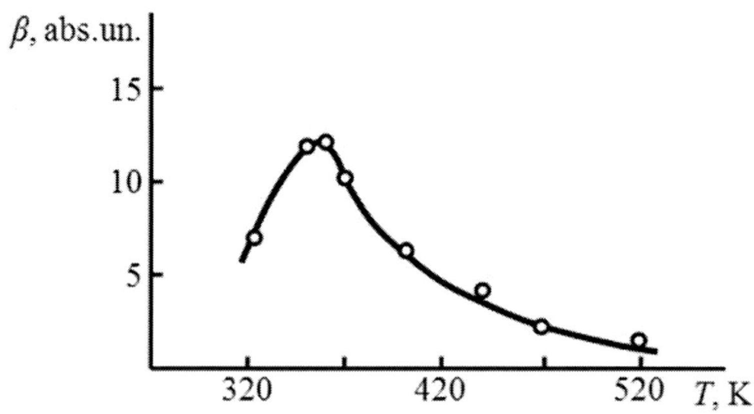

Figure 5.64. Dependence of the sensitivity to the sulfur dioxide on the operating temperature of the sensor based on the cadmium sulfide film.

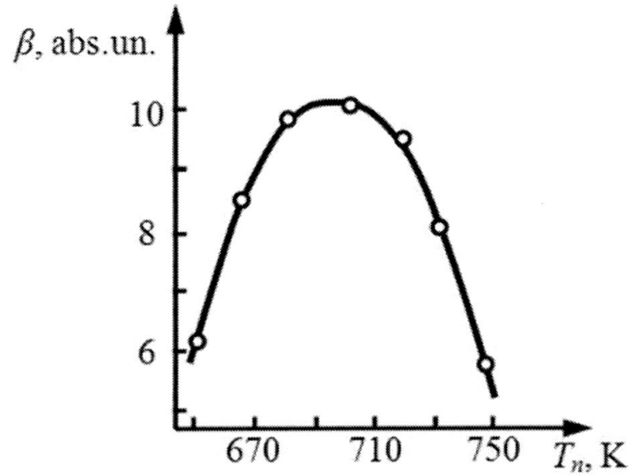

Figure 5.65. Dependence of sensitivity β to the SO_2 on the substrate temperature, at which applied the *CdS* sensitive layer of the sensor.

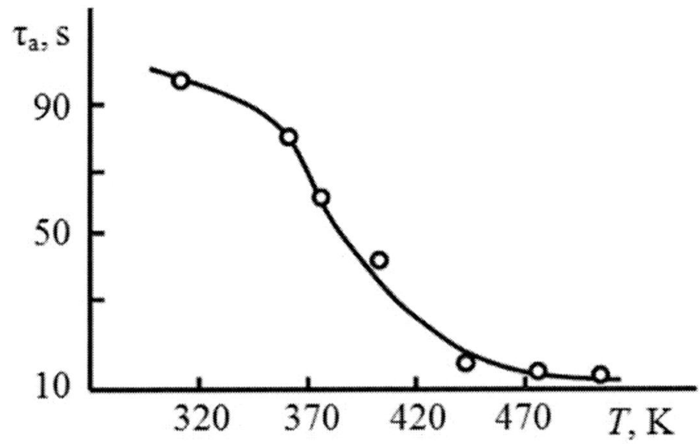

Figure 5.66. Dependence of the sulfur dioxide adsorption time on the temperature of the sensor based on the cadmium sulfide film.

Characteristically, the samples with differing nature of the major donors in the surface area, the maximum on the curve β (T) achieves at different temperatures [29, 30]. The maximum value β of the samples, the major donors of which is the selenium vacancies, observed at 460 *K*. At *CdSe* sensors films with major donors – the interstitial cadmium atoms, the β maximum founds at $T \approx 410$ *K*. As shown in [30], these *CdSe* sensors films are different by the main form of chemisorbed oxygen on the surface. In the first case, the main form of chemisorption is atomic oxygen, in the second - O_2. Consequently, the position of the maximum on the curve β (T) depends on the composition of the sensor surface, which defines the relationship between the forms of chemisorbed oxygen.

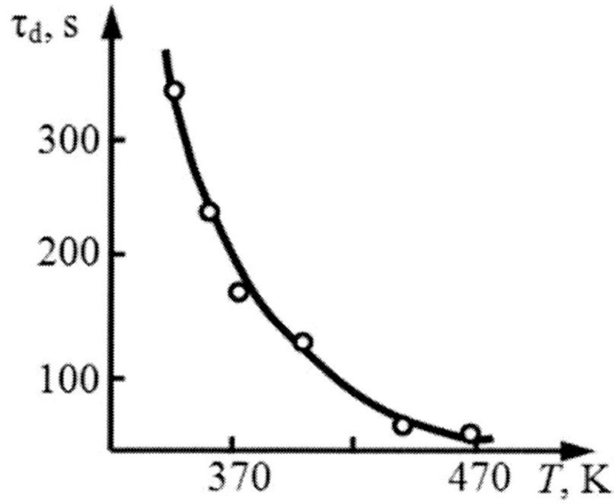

Figure 5.67. Influence of sensor temperature, that based on the cadmium sulfide film, on the sulfur dioxide desorption time.

5.11.2. Selectivity of the Gas Sensors

Selectivity problem of the gas sensors is one of the most pressing and relevant in the analytical instrument-making, especially in an industry, that is based on the semiconductor sensors.

Operation of the most semiconductor gas sensors is based on the change of its macro parameters, which describes and characterizes the collective interactions on the surface (conductivity, photo-conductivity, surface potential barrier, photo-electromotive force, capacity, mobility, and concentration of free charge carriers, and others).

The main process on the surface, which leads to the changes of the macro parameters, is the interaction of adsorbing gas particle to the adsorption center on the surface. In this case, the main parameters, that determine the course of the interaction between it, is the temperature and difference between the values of the chemisorbed particle electron affinity and adsorption center.

Temperature determines the charge state of the energy level in the band gap of the defect on the surface of the adsorption center. Of the value of electro-negativity of the gas particle depends, whether at the given temperature its reaction of the semiconductor being the donor or the acceptor.

Therefore, under these conditions, a semiconductor gas sensor β is essentially able to distinguish two classes of the gases: donor and acceptor. Even within a single class of gases, the question of the selectivity becomes a very complicated, and in it solving presents the fundamental problems. In part, they may be solved by doping the semiconductor sensor surface, for example, by the atoms of various metals, that creates in the band gap the energy levels with different energies from the bottom of the conduction band. In this case, for the gas particles that differ in value of the electro-negativity, creates the adsorption centers, that are more suitable for interaction. However, all of these interactions are reproducing in the same of the above parameters. This leads to interference of the signals and makes the difficult the selection process of it's the desired characteristics.

By selecting the required operating temperature, it is possible to create on the surface the conditions, at which of the two, for example, the acceptor gases, by a significant difference in the values of electro-negativity, the chemisorption of the more electronegative gas is possible, and the gas with a lower electro-negativity is already in the desorption state. In such circumstances, managed to register a mini-SO_2 concentration in the normal air atmosphere, i. e. at the presence of macroscopic oxygen concentration [30, 31].

In some cases, the problem of selectivity solving by creating a matrix of semiconductor gas sensors. In the matrix may be present the same type semiconductor gas sensors or different by the chemical composition. But the main difference between sensors - it's operating temperature. This is caused by the fact that in most cases, the interaction between semiconductor gas sensors and gas particle is determined by electronegativity of the particle. In this case, the sensors surfaces of the matrix elements may be doping by the atoms of various materials in order to increase the sensor selectivity, or the matrix as a whole to one particular class of gas or gases.

In the case of doping the surface by atoms' ligands, it is fundamentally important, which bond will be formed at the ligand atom orbitals at the interaction with chemisorbed particle: binding, anti-binding, not binding. Atom of the ligand is noticeably sensitive to the chemisorption of gas particles, if only the bonding orbital is formed. In this case, the sensitivity to the presence of the adsorbed particles on the surface can be the macro-electrical sensor parameter. These parameters characterize the collective phenomena of the sensors and could be easily measured. Methods of the measurement of these parameters (*σ*, *μ*, *n*, *γ*, etc.) allow the sensor miniaturization and its articulation with the micro-electronic devices. These devices could be operated in the remote observation points in stand-alone mode. Interaction of adsorption centers on the surface with adsorbed particle could possess a local character. In this case, the sensor macro-electrical parameters remain virtually unchanged. However, it significantly changes the absorption and emission atomic spectra of the adsorption centers and adsorbed particle. In this case, applying the optical methods for analyzing the changes in the intensities of the lines of the absorption and emission spectra of the atoms, which are the parts of the adsorbed complex on the surface (adsorption center - adsorbed particle). In many cases, the optical methods are more accurate, it provides the desired selectivity and sensitivity, minimal drift of the usable signal. However, the optical systems are generally quite bulky and cannot be used primarily for operate in the standalone mode at the remote locations.

The intensive use at the last decades, the optical fiber to transmit the optical signal, partially solves the indicated problems of the optical sensor systems. Particularly attractive is the using of the fiber optics devices in the combination with optical nano-sensors [93-96], which not requires the electrical power and contacts, and the optical signal through an optical fiber can be transmitted over long distances.

REFERENCES

[1] Lantto, V; Golovanov, V; Lychkovsky, Y. Carbon Monoxide Monitoring in Combustion Emissions Using SnO_2 and In_2O_3 Gas Sensors. In: *The Proc. TTFS'94 Workshop.* Szklarsla Poreba, Poland: 1994, 180-183.

[2] Malchenko, SN; Lychkovsky, YN; Baykov, MN. One-electrode semiconductor sensor for detection of toxic and explosive gases in air. *Sensors and Actuators B*, 1992 vol. 7, 505–506.

[3] Malchenko, SN; Lychkovsky, YN; Baykov, MN. In_2O_3 based gas sensors. *Sensors and Actuators B*, 1993 vol. 13–14, 159–161.

[4] Baran, SV; Branitsky, GA; Ivanovskaya, MI. Thermocatalytic sensors with Pd-Pt- Al_2O_3 catalyst. *Sensors and Actuators B*, 1993 vol.13–14, 244–247.

[5] Orlik, DR; Ivanovskaya, MI; Branitsky, GA; Bogdanov, PA. Monoelectrode gas sensors based on SnO_2 semiconductor ceramics. *Sensors and Actuators B*, 1993 vol.13–14, 605–607.

[6] Yasukawa, Y; Seki, T; Jun-Ichi Muramatsu. Film-type In_2O_3 gas sensors. *Sensors and Actuators B*, 1993 vol.13–14, 613–614.

[7] Matsuura, S. New developments and applications of gas sensors in Japan. Sensors and *Actuators B*, 1993 vol.13–14, 7–11.

[8] Голованов, ВВ. Характеризация полупроводниковых поверхностно-активных веществ $Sn_xW_yO_z$, $Cd_xS(Se)_{1-x}$, In_2O_3. *Дис. ... д-ра физ.-мат. наук*. Научный консультант Смынтына, ВА.. Одесса: ОГУ; 1998.

[9] Shimada, M;. Watanabe, H; Ishida, T. Manufacturing of fineline films by printing technique In: *Proc. IMC 1990*. Tokyo: 1990, 581–586.

[10] Nakao, K; Hayama, M; Karamasu, K; [and others]. Giant Micro-electronics technology using gravure printing techniques. In: *Extended Abstract of the 1991 International Conference on Solid State Devises and Materials*. Yokohama:1991, 599-601.

[11] Stein, G; Sidney, J; Richard, L; [and others]. Interconnection and packaging of advanced electronic circuitry. In: *Proc. ISHM Conference*.Florida: 1991.

[12] Leppavouri, S; Vaananen, J; [and others]. A novel thick film technique, gravure offset printing, for the realisation of fine line sensor structures. In: *The Proc. Eurosensors Conf.* Budapest:1993, 123-125.

[13] Barret, J; Morgan, P. Laser drilling of vias in unfired thick film dielectrics. In: *Proc. ISHM Conf.* Dallas:1984, 1-7.

[14] Hirakawa, T; Kato, C. The application of eximer laser to fine line feature interconnections. In: *The Proc. IMC*. Yokohama: 1992, 56-59.

[15] Haertling, C; Yoshikawa,S; Newnham, R. Thick film patterned ceramics using UV-curable pastes. *J. Amer. Ceramic Soc.*, 1990 vol. 73 no. 11, 3339–3344.

[16] Felten, J; Collins, J. Small via generation by diffusion pattering.In: *The Proc. ISHM Conf.* 1990, 224-227.

[17] Коженков, ВН; Фукс, НА. Электрогидродинамическое распыление жидкости .*Успехи химии*, 1976 т. 45, 2274–2283.

[18] Marrucchi, J; Perotin, M; [and others].Influence of heat treatment on the properties of CdS sprayed films. First results on Cu_xS sprayed films. In: *Conr. Rec. 13th IEEE Photovolt. Spec. Conf.*,Wash., 1978, 298-302

[19] Колпаков, АВ; Контуш, СМ; [и др.]. Метод получения фоточувствительных плёнок сульфида кадмия .*Авторское свидетельство СССР №890907*. Опубл. БИ № 10, 1982.

[20] Авербух, ЕМ; Семенов, ВН. О получении легированных плёнок методом пульверизации. В: *Физико-химия полупроводникового материаловедения*. Воронеж: Изд. ВГУ; 1979, 100-103.

[21] Голованов, ВВ; Кальницкий, ВГ; [и др.]. Способ и устройство для получения легированных полупроводниковых плёнок методом электрогидродинамического распыления жидкости. *Авторское свидетельство СССР № 1366000.*

[22] Yale, M; Fahrenburch,AL; Bube, RH. Photovoltaic properties of n-CdS / p-CdTe heterojunctions prepeared by spray pyrolisis. *Appl. Phys. Letters*, 1977 vol. 30, no. 8,. 423–424.

[23] Alaee, MS; Rouhani, MD. Preparation and characterization of sprayed CdS films. In: *The Proc Photovolt. Solar Energy Conf.* Luxembourg: 1997, 1151-1159.

[24] Прокатор, ЛМ; Еремчук, ЛП; Баско, ОА. Фоточувствительные слои селенида кадмия, приготовленные способом химического распыления *Электронная техника. Электроннолучевые и фотоэлектрические приборы*, 1970 №1, 51.

[25] Алхазов, ТГ; Марголис, ЛЯ. *Глубокое каталитическое окисление органических материалов*. М. : Химия; 1985.

[26] Григорьев, ВГ. Метод криодисперсии: состояние и перспективы. *Вестник академии наук СССР*, 1987 №4, 84–90.

[27] Сердюк, ВВ; Смынтына, ВА. Разработка и исследование измерительного преобразователя для определения концентрации кислорода в газовых смесях . *Отчет о НИР*; Одесса: Одес. гос. ун-т; 1978.

[28] Сердюк, ВВ; Смынтына, ВА. Влияние отжига на воздухе на структуру, химический состав и электрические свойства тонких слоев селенида кадмия. *Полупроводниковая техника и микроэлектроника*, 1976 вып. 22, 5-8.

[29] Вашпанов, ЮА; Смынтына, ВА. *Адсорбционная чувствительность полупроводников.* Одесса : Астропринт; 2005.

[30] Smyntyna V. *Electron and Molecular Phenomena on the Surface of Semiconductors*. New York: NOVA Publishers; 2013.

[31] Смынтына, ВА. Электрофизические, фотоэлектрические и электронно-молекулярные явления, стимулированные хемосорбционно-диффузионными процессами на границах полупроводниковых поликристаллических пленок селенида и сульфида кадмия. *Дис. д-ра физ.- мат. наук*. Одесса : ОГУ; 1988.

[32] Вашпанов, ЮА; . Проничкин, ВД; Смынтына, ВА; Сердюк, ВВ. Чувствительный элемент газоанализатора на кислород. *Авторское свидетельство СССР. №1106583.*

[33] Смынтына, ВА; Вашпанов, ЮА; Сердюк, ВВ. Чувствительный элемент газоанализатора. *Авторское свидетельство СССР №1151082.*

[34] Мартинуцци, С. Свойства плёнок CdS, полученных методом реактивного распыления . В:Солнечная энергетика. М: Мир; 1979, 325–331.

[35] Нестеренко, ЮА; Середенко, АС; Смирнова, ОА. К вопросу о термодинамике образования кислородосодержащих примесей в соединениях A_2B_6 В: *Сборник научных трудов ВНИИ монокристаллов и синтиляционных материалов и особо чистых веществ*. Харьков:1979 № 3, 113–117.

[36] Вашпанов, ЮА. Адсорбционная чувствительность полупроводниковых материалов A_2 B_6, оксидов тяжелых металлов и кремния с реальной поверхностью с кластерными структурами : *Дис. д-ра физ.-мат. наук*. Научный консультант Смынтына, ВА. Одесса: ОГУ;1999.

[37] Стефаняк, АВ.; Михеева, ИЛ. *Методы и приборы для анализа сернистого газа В: Элементы, устройства и системы газового анализа*. К.: Наукова думка;1979, 3–7.

[38] Сморчков, ВИ. Приборы газового анализа на основе использования адсорбционных свойств поверхности полупроводников. В: *Современные инструментальные методы и средства газового анализатора.* К: Изд. ВНИИАП; 1985, 9–14.

[39] Голованов, ВВ; Павленко, ГК; Смынтына, ВА; Шмилевич, АМ. Исследование возможности создания газоанализатора на основе полупроводниковой плёнки сульфида кадмия. В: *Физико-химические методы и инженерно-технические решения в газоаналитическом приборостроении.* К: Наукова думка; 1984, 110–111.

[40] Голованов, ВВ; Чемересюк, ГГ; Шмилевич, АМ. Модуляция потенциального рельефа и изменение природы собственных дефектов CdS в результате термообработки В: *Материаловедение халькогенидных и кислородосодержащих полупроводников. Тез. докладов Второй Всесоюзной научно-технической конференции.* Черновцы; 1986 т.1, 178.

[41] Голованов, ВВ; Сердюк, ВВ; [и др.] Температурная зависимость фототока и её связь с электронно-молекулярными процессами на поверхности плёнок сульфида кадмия. В: *Физика поверхностных явлений в полупроводниках*. К: Наукова думка; 1984,. 86–87.

[42] Голованов, ВВ; Смынтына, ВА; Чемересюк, ГГ; Шмилевич, АМ. Адсорбционно-чувствительный к SO_2 элемент на основе сульфида кадмия. В: *Современные инструментальные методы и средства газового анализа.* К : ВНИИАП; 1985, 20–22.

[43] Голованов, ВВ; Чемересюк, ГГ; Шмилевич, АМ. Адсорбция и диффузия кислорода в тонких поликристаллических плёнках сульфида кадмия. *Журнал Физической Химии*, 1986 т.60, 2881–2882.

[44] Сердюк, ВВ; Смынтына, ВА. Разработка методов повышения качества полупроводниковых чувствительных элементов на SO_2 и анализ возможности создания датчиков на H_2. *Отчет о НИР*. Одесса: ОГУ; 1990.

[45] Киселёв, ВФ; Крылов, ОВ. *Электронные явления в адсорбции и катализе на полупроводниках и диэлектриках.* М: Наука; 1979.

[46] Morrison, SR. *The Chemical Physics of Surfaces*. New-York and London: Plenum Press; 1977.

[47] Дубовой, ВК; Кучеров, НЯ; Шияновский, ВИ. Исследование некоторых электрических свойств поверхности монокристаллов CdS. *Украинский Физический Журнал*, 1966 т.11, 383–388.

[48] Смынтына, ВА. Повышение и стабилизация адсорбционной способности поверхности плёнок CdSe. В: *Новые физические принципы в аналитическом приборостроении*. К.: Наукова думка, 1980, 35–37.

[49] Ян де Бур. *Динамический характер адсорбции* М.: Иностр. лит.; 1962.

[50] Sberverglieri G. Classical and novel techniques for the preparation of SnO_2 thin-film gas sensors. *Sensors and Actuators B*, 1992 vol. 6, 239–247.

[51] Fleischer M. Meixner, H. Sensing Reducing gases at high temperatures using long-term stable Ga_2O_3 thin-films. *Sensors and Actuators B*, 1992 vol. 6, 257–261.

[52] Martinelli, G; Carotta, MC. Sensitivity to reducing gas as a function of energy barrier in SnO_2 thin films. *Sensors and Actuators B*, 1992 vol.7, 717–720.

[53] Tragut C. The influence of the surface transfer reaction on the response characteristics of resistive oxygen sensors. *Sensors and Actuators B*, 1992 vol. 7, 742–746.

[54] Sberverglieri, G; Gropelli, S; [and others]. Oxygen gas sensing characteristics for ZnO(Li) sputtered thin films. *Sensors and Actuators B*, 1992 vol. 7, 747–751.

[55] Chu, WF; Leonard, V; [and others]. Thick film chemical sensors. *Sensors and Actuators B,* 1992 vol. 7, 321–324.

[56] Gardner, JW; Shurmer, HV; Cocoran, P. Integrated tin oxide odour sensors. *Sensors and Actuators B*, 1991 vol. 4, 117–121.

[57] Gerblinger, J.; Meixner, H. Fast oxygen sensors based on sputtered strontium titanate .*Sensors and Actuators B*, 1991 vol. 4, 99–102.

[58] Morrison, SR. Semiconducting Gas Sensors. *Sensor and Actuators.* 1982. vol. 2, 329–341.

[59] Geatches, RM; Chadwich, AV; Wright, JD. Single-crystal metal oxide gas sensors. *Sensors and Actuators B*, 1991 vol. 4, 467–462.

[60] Grinevich, VS; Smyntyna, VA. Electronic mechanism for adsorptive sensitivity in semi-conductor gas sensor. *Sensors and Actuators B*, 1994 vol. 18–19, 426–428.

[61] Honore, M; Lenaerts, S; [and others]. Synthesis and characterization of tin dioxide powders for the realization of thick film gas sensors. *Sensors and Actuators B*, 1994 vol.18-19, 621–624.

[62] Reti, F;Fleischer, M; [and others]. Effect of coadsorption of reducing gases on the conductivity on -Ga_2O_3 thin films in the presence. *Sensors and Actuators B*, 1994 vol. 18-19, 573–577.

[63] Gerblinger, J; Lampe, U; [and others]. Cross sensitivity of various doped strontium titanate films to CO, CO_2, H_2, H_2O and CH_4. *Sensors and Actuators B*, 1994 vol. 18–19, 529–534.

[64] Fliegel, W; Behr, G; [and others]. Preparation, development of microstructure, electrical and gas –sensitive properties of pure and doped SnO_2 powders. *Sensors and Actuators B,* 1994 vol. 18-19, 474–477.

[65] Andreev, SK; Popova, LI; [and others]. Characteristics of diode humidity sensors based on SnO_2 sensing properties. *Sensors and Actuators B*, 1994 vol. 18–19, 457–459.

[66] Geistlinger, H. Accumulation layer model for Ga_2O_3 thin-film gas sensors based on the Volkenstein theory of catalysis. *Sensors and Actuators B*, 1994 vol. 18–19, 125–131.

[67] Göppel, W. New materials and transducers for chemical sensors. *Sensors and Actuators B,* 1994 vol. 18–19, 1–21.

[68] Хедвалл, ДА. *Химия твёрдого состояния*. Токио: Каиау донинж 1968.

[69] Татио, К. Газовый детектор с применением тонкой плёнки фталоциамина. *Дэнки кадаки*, 1979. т. 46, № 11, 597–603.

[70] Веденеев, ВИ. *Энергия разрыва химических связей. Потенциалы ионизации и сродство к электрону. Справочник*. М.: Изд. АН СССР; 1962.

[71] Гороновский, ИТ; Назаренко, ЮП; Некриач; ЕФ; под ред. Пилипенко, АТ. *Краткий справочник по химии*. 5-е изд. К.: Наукова думка; 1987.

[72] Волькенштейн, ФФ. *Электронные процессы на поверхности полупроводников при хемосорбции*. М.: Наука; 1987.

[73] Голованов, ВВ; Курков, ВВ; Смынтына, ВА Способ получения чувствительного элемента анализатора окиси углерода. *Авторское свидетельство СССР №1775657*.

[74] Голованов, ВВ; Смынтына, ВА; Г. Г. Чемересюк, ГГ; Шмилевич, АМ. Чувствительный элемент газоанализатора сернистого ангидрида. *Авторское свидетельство СССР № 1345103*.

[75] Смынтына, ВА. Исследование электронно-молекулярных явлений на поверхности полупроводниковых слоев и создание научных основ разработки сенсоров для регистрации газовой среды. *Отчет о НИР*. № госрегистрации 0197U008970. Одесса: ОГУ; 1999.

[76] Cross, YM.;. Pyke, DR. An X-ray photoelectron spectroscopy study of the surface composition of tin and antimony mixed metal oxide catalysis. *Journal of catalysis*, 1979 vol.58, 61–67.

[77] Смынтына, ВА. Исследование электронно-молекулярных явлений на поверхности полупроводниковых слоев и разработка научных основ физико-химических процессов в полупроводниковых сенсорах. *Отчет о НИР*. Одесса: Одесский национальный ун-т им. И.И.Мечникова; 2002.

[78] Голованов, ВВ; Смынтына, ВА; Гудис, А. Полупроводниковый сенсор на основе CdS. *Журнал аналитической химии*, 1991 т. 46 2374–2379.

[79] Solis, JL;. Golovanov, VV;. Lantto, V. A study of dual conductance response to carbon monoxide of CdS and $SnWO_4$ thin films, *Physica Scripta*, 1994 vol. 54, 548–251.

[80] Chiotti, G; Boccuzzi, F; Siorino, A. Surface characterization of Cu/ZnO catalysis: IR, VIS, UV Study of CO chemisorptions. In: *Bond*

Adsorption and Catalysis on Oxide Surfaces. Amsterdam: Elsevier Science Publishers B. V.; 1985.

[81] Сердюк, ВВ; Смынтына, ВА. Разработка и исследование измерительного преобразователя для определения концентрации кислорода в газовых смесях. *Отчет о НИР*. № госрегистрации 77075859. Одесса: ОГУ; 1978.

[82] Мясников, ИА. Об особенностях адсорбции молекул кислорода и радикалов на поверхности окислов металлов в газовых и жидких средах. *Журнал физической химии*, 1981 т.55, 1278–1282.

[83] Roberts, MW; Ross, RH. Sulphidizing kinetics of lead and iron films. *Trans. Faraday Soc*., 1966 vol. 62, 2301–2304.

[84] Сердюк, ВВ; Смынтына, ВА. Разработка технологии изготовления датчиков концентрации кислорода, двуокиси серы и сероводорода. *Отчет о НИР*. № госрегистрации 01840019476. Одесса: ОГУ; 1984.

[85] Takahata, K. Tin dioxide sensors. Development and applications. In: *Chemical Sensor Technology*. 1988, vol. 1, 39-45.

[86] Lantto, V; Mizsei, J. H_2S monitoring as an air pollutant with silver doped $Sn0_2$ thin-film sensors. In: *Proc. 3rd Int. Meet. Chemical Sensors*. Cleveland, OH: 1990, 39–43.

[87] Rappe, AK; Goddard, WA. Charge equilibration for molecular dynamics simulations. *J. Phys. Chem*., 1991 vol. 95, 3358–3363.

[88] Rantala, TT. *Ab initio* Studies of Compound Semiconductor surfaces. In: *Computational Studies of New Materials*. Singapore: World Scientific; 1999, 6-26.

[89] Голованов, ВВ; Сердюк, ВВ. Механизм хемосорбции кислорода на поверхности поликристаллических плёнок CdS. *Украинский физический журнал*, 1988 т.33, 157–162.

[90] Маркова, ВН. Полевые эффекты на границах зерен многофазной поликристаллической структуры чувствительного слоя адсорбционного элемента как фактор, определяющий чувствительность к газам: В: *Сб. научных трудов ВНИИАП. Анализ газов с применением современных средств микроэлектроники и полупроводниковой техники*. К: ВНИИАП; 1982, 33–41.

[91] Рыльцев, ИВ; Гутман, ЭЕ; Мясников ИА. Влияние “эффекта поля” на адсорбцию атомов и молекул водорода и кислорода на окиси цинка. *Журнал . физической химии*, 1981 т. 55, 968–976.

[92] Волькенштейн, ФФ. Физико-химия поверхности полупроводников. М.: Наука; 1973.

[93] Skobeeva, VM; Smyntyna, VA; Sviridova, OV; Struts, DA; Turin, AV. Optical properties of Cadmium Sulphide Nanocrystals obtained by Sol-Gel Method in Gelatine, *Journal of Applied Spectroscopy*, 2008 vol. 75, 576–582.

[94] Smyntyna, V; Pisco,M; Consales, M; Viter, R. Novel Optochemical Sensor based on SnO_2 Sensitive Thin Film for ppm Ammonia Detection in Liquid Environment. *Journal of light wave technology*, 2006 vol.24, 5000–5007.

[95] Smyntyna, V; Skobeeva, V; Malushin, N. The nature of emission centers in CdS nanocrystals. *Journal of Radiation Measuraments*, 2007 vol. 42, 693–696.

[96] Smyntyna, V; Cusano, A; Consales, M; Viter, R. Optochemical sensor for water monitoring based on SnO_2 particle layer deposited onto optical fibers by the electroscopy pyrolysis. *Applied Physics Letters*, 2006 vol. 89, 3484-3486.

ABOUT THE AUTHOR

Valentyn SMYNTYNA - Full Professor, Doctor of Sciences in Physics and Mathematics, Director of Experimental Physics Department of Odessa National I.I. Mechnikov University, Head of Scientific and Technical Center and Research Laboratory of electron and molecular processes on semiconductor surfaces. Honoured Scientist of Ukraine (2003), Laureate of State Premium of Ukraine in the field of Science and Technology (2007).

Principal Areas of Scientific expertise

Physics of Semiconductors, Surface Physics, Sensorics. Main scientific results have been achieved in the field of experimental physics, physics of surface effects in the course of investigation of surfaces of films, layers, composite macro-, micro- and nanoporous structures and quantum dots of semiconductors. He has solved principally important physical problem of purposeful influence on surface adsorptive properties; has developed theory of universal definition of adsorptive sensitivity of materials, methods of its prognostics and formation. He is the first one who has proposed the method of elemental diagnostic of surface composition, which is based on results of interaction among proper and adsorbed atoms of surface. He has detected new type of heterogeneity on a surface - chemisorptional-electric domain - responsible for complex of surface phenomena which have been studied for the first time: chemisorptionally stimulated fluctuations of current, negative differential resistance and saturation of volt-ampere characteristics, sensibilization and desensibilization of surface in the result of chemisoptional generation and decomposition of centres of photosensitivity.

Most important results achieved under scientific supervision of Valentyn Smyntyna are discussed as 9 Doctor of Sciences and 19 Candidate of Science (PhD) Dissertations, were marked by medals at International and Pan-Ukrainian exhibitions.

Publications

He is the author of about 700 published works, among which there are 11 scientific monographs (5 of them are personal), over 200 scientific papers in leading international and domestic editions (i.e. Applied Physics Letters, Journal of light wave technology, Radiation Measurements, Journal of Applied Spectroscopy, Semiconductors, Thin Solid Films, Technical Physics Letters, Vuoto, Russian Physics Journal, Russian Journal of Physical chemistry, II Nuovo Cimento A, Physica Status Solidi (a), ect.), 24 patents (7 personal), and 11 textbooks (5 personal). Most important of these publications are cited in the reference section of this book and in the monograph 'Valentyn Smyntyna. Electron and Molecular Phenomena on the Surface of Semiconductors. NOVA Publishers, 2013'.

Courses Taught

Optics, Photoelectric Phenomena in Semiconductors, Physical-Chemical Phenomena on the Surface of Solid States.

Professional Membership

American Physical Society, Optical Society of America, American Chemical Society, European Physical Society, Ukrainian Physical Society – vice-president, Academy of Higher Education of Ukraine – full member, EUROSENSORS International Steering Committee – honorary member.

Public Activity

Member of Committee of State Awards of Ukraine in the field of science and technology, first deputy head of Southern Scientific Centre of National Academy of Science of Ukraine and Ministry of Education and Science of Ukraine, vice-head of Scientific Council for Physics of Semiconductors of Presidium of National Academy of Science of Ukraine, President of Specialized Scientific Council for dissertations of Candidate (PhD) and Doctor (Dr.Sc.) of Physical and Mathematical Sciences Degree at Odessa National I.I. Mechnikov University, editor in chief of scientific magazines "Sensor electronics and microsystem technologies", "Photoelectronics".

Honours and Awards

Laureate of State and non-governmental Orders, Medals, Premiums, Honours and Awards of Ukraine, Italy, Poland, Vatican, USA, UK, France, Finland, Turkey; Senator *Honores Causae* of Szeged University (Hungary), Professor *Honores Causae* of Moskow Humanitarian University (Russia), Kherson National Technical University (Ukraine), Galysh Academy (Ukraine), Academy of Vatican, Qingdao University (China), Transnistrian University (Moldova); Honorary citizen (freeman) of the city of Ochakiv (Ukraine).

INDEX

B

C

D

E

F

G

H

I

J

K

L

M

N

O

P

T